T0205966

# SpringerBriefs in Fire

*Series Editor*

James A. Milke

More information about this series at http://www.springer.com/series/10476

Nathan White • Michael Delichatsios

# Fire Hazards of Exterior Wall Assemblies Containing Combustible Components

Nathan White
CSIRO
Highett, VIC, Australia

Michael Delichatsios
FireSERT
University of Ulster
Jordanstown, Ireland

ISSN 2193-6595          ISSN 2193-6609   (electronic)
SpringerBriefs in Fire
ISBN 978-1-4939-2897-2          ISBN 978-1-4939-2898-9   (eBook)
DOI 10.1007/978-1-4939-2898-9

Library of Congress Control Number: 2015944257

Springer New York Heidelberg Dordrecht London
© Fire Protection Research Foundation 2015
This work is subject to copyright. All rights are reserved by the Publisher, whether the whole or part of
the material is concerned, specifically the rights of translation, reprinting, reuse of illustrations, recitation,
broadcasting, reproduction on microfilms or in any other physical way, and transmission or information
storage and retrieval, electronic adaptation, computer software, or by similar or dissimilar methodology
now known or hereafter developed.
The use of general descriptive names, registered names, trademarks, service marks, etc. in this publication
does not imply, even in the absence of a specific statement, that such names are exempt from the relevant
protective laws and regulations and therefore free for general use.
The publisher, the authors and the editors are safe to assume that the advice and information in this book
are believed to be true and accurate at the date of publication. Neither the publisher nor the authors or the
editors give a warranty, express or implied, with respect to the material contained herein or for any errors
or omissions that may have been made.

Printed on acid-free paper

Springer Science+Business Media New York is part of Springer Science+Business Media
(www.springer.com)

# Foreword

Many combustible materials are used today in commercial wall assemblies to improve energy performance, reduce water and air infiltration, and allow for aesthetic design flexibility. These assemblies include Exterior Insulation Finish Systems (EIFS), metal composite claddings, high-pressure laminates, and weather-resistive barriers (WRB). The combustibility of the assembly components directly impacts the fire hazard. For example, the insulation component of EIFS and other emerging-related systems (e.g., Structural Insulation Finish Systems (SIFS)) is combustible foam which exhibits rapid flame spread upon fire exposure. There have been a number of documented fire incidents involving combustible exterior walls but a better understanding was needed of the specific scenarios leading to these incidents to inform current test methods and potential mitigating strategies.

The Foundation initiated a project with an overall goal to develop the technical basis for fire mitigation strategies for fires involving exterior wall systems with combustible components. The goal of this first phase project is to compile information on typical fire scenarios which involve the exterior wall, compile relevant test methods and listing criteria as well as other approval/regulatory requirements for these systems, and to identify the knowledge gaps and the recommended fire scenarios and testing approach for possible future work.

Highett, VIC, Australia                                            Nathan White
Jordanstown, Ireland                                    Michael Delichatsios

Foreword

# Preface

It is recognized from past experience that fire spread from floor to floor and over the façade in buildings can be a catastrophic event. The regulatory and test-based methodology to address behaviour of fires in facades for different facade systems varies significantly for different countries. One difficulty is that several materials and assemblies are involved, such as Timber, Plastics, GRP, Glazing, Polymeric composites, Cement-based products, with and without insulation. The other difficulty is that there is not a consensus to select the size of exposure fire for testing and evaluation of any given facade system. The fire hazard is more severe if components of the facade assembly are combustible. Some past fire incidents have demonstrated rapid and extensive fire spread over the length of the façade either externally or internally through the insulation cavity. For facades with no combustible components, fire spread may occur from floor to floor in a leap frogging fashion.

Phase I of this study seeks and is structured to collect data about combustible facade systems, review existing research in this area, examine statistics on façade fires, list incident of facade fires, describe the mechanisms and dynamics of fire spread, review existing test methods and performance criteria, and conclude with recommendations for a testing approach and methodology for a possible future experimental research Phase II.

This project has been prepared by CSIRO and FireSERT (university of Ulster) for the Fire Protection Research Foundation (FPRF) as the deliverable for the project *Fire Hazards of Exterior Wall Assemblies Containing Combustible Components— Phase I*. This phase investigated the following items:

- Combustible exterior wall systems in common use
- Existing research and mechanisms of fire spread
- Fire statistics
- Fire incident case studies
- Test methods and regulations

Types of combustible exterior wall systems in common use include:

- Exterior Insulation Finish Systems (EIFS, ETICS, or synthetic stucco)
- Metal composite material cladding (MCM)
- High-pressure laminates
- Structural Insulation Panel Systems (SIPS) and insulated sandwich panel systems
- Rain screen cladding or ventilated facades (curtain walls)
- Weather-resistive barriers (WRB)
- External timber panelling

These exterior wall systems are typically complex assemblies of different material types and layers which may include insulation layers and vertical cavities, with or without fire stopping.

A brief overview of existing research related to fire performance of exterior combustible walls is provided. The Fire Code Reform Centre funded a research report on fire performance of exterior claddings [1] that provides an excellent overview of the previous research up to the year 2000. Appendix D also provides a list of related research literature for further reading.

The key initiating fire can be one of the two possible types of fires:

I. Fires external to the building (other burning buildings, external ground fires) or
II. Fires internal to the building originating in a floor that have resulted in breaking the windows and ejecting flames on the façade

Key mechanisms of fire spread after initiating event include:

I. Fire spread to the interior of level above via openings such as windows causing secondary interior fires on levels above resulting in level to level fire spread (leap frogging).
II. Flame spread over the external surface of the wall if combustible.
III. Flame spread within an interval vertical cavity/air gap or internal insulation layer. This may include possible failure of any fire barriers if present, particularly at the junction of the floor with the external wall.
IV. Heat flux impacts causing degradation/separation of non-combustible external skin (loss of integrity) resulting on flame spread on internal core.
V. Secondary external fires to lower (ground) levels arising from falling burning debris or downward fire spread.

It follows from this research review that the façade fire safety problem can be divided into four parts:

1. Specification of fire exposure scenario and the heat flux distribution both inside the enclosure and from the façade flames originating from the fire in the enclosure. This requirement is prerequisite for the following parts.
2. Fire resistance of the façade assembly and façade-floor slab junction including structural failure for non-combustible and combustible façade assemblies.

3. Fire spread on the external surface of the façade assembly if combustible due to the flames from the enclosure fire.
4. Fire spread and propagation inside the façade insulation, if combustible, due to the enclosure fire.

Statistics relating to exterior wall fires from the USA, Australia, New Zealand, and Nordic countries have been reviewed. Statistical data relating to exterior wall fires is very limited and does not capture information such as the type of exterior wall material involved, the extent of fire spread, or the mechanism of fire spread. Exterior wall fires appear to account for somewhere between 1.3 and 3 % of the total structure fires for all selected property types investigated. However, for some individual property types, exterior wall fires appear to account for a higher proportion of the structure fires, the highest being 10 % for storage type properties. This indicates that exterior wall fires are generally low-frequency events, particularly compared to fires involving predominantly the interior. The statistics also indicate that sprinkler systems are likely to have an effect on the risk of exterior wall fires by reducing the risk of spread from an internal fire to the exterior façade. However, a significant portion of external wall fires still occur in sprinkler protected buildings, which may be due to both external fire sources and/or failure of sprinklers. On this basis, it is recommended that controls on flammability of exterior wall assemblies should be the same for sprinkler protected and non-sprinkler protected buildings.

Fire incidents involving exterior wall assemblies around the world have been reviewed. This review indicates that although exterior wall fires are low-frequency events, the resulting consequences in terms of extent of fire spread and property loss can be potentially very high. For most of the incidents reviewed, the impact on life safety in terms of deaths has been relatively low with the main impacts being due to smoke exposure rather than direct flame or heat exposure. However, a large number of occupants are usually displaced for significant periods after the fire incidents. This has particularly been the case for incidents in countries with poor (or no) regulatory controls on combustible exterior walls or where construction has not been accordance with regulatory controls. Combustible exterior wall systems may present an increased fire hazard during installation and construction prior to complete finishing and protection of the systems. The 2009 CCTV Tower Fire and 2010 Shanghai fire in China are examples of large fires occurring during construction.

Regulation and building code requirements for fire performance of exterior wall assemblies around the world have been reviewed. Five aspects of regulation have been identified to influence the risk of fire spread on exterior wall systems. These include reaction to fire of exterior wall systems and individual components, fire stopping of cavities and gaps, separation of buildings, separation of openings vertically between stories of fire compartments and sprinkler protection. Of these, the reaction to fire regulation requirements are expected to have the most significant impact on actual fire performance and level of fire risk presented by exterior wall assemblies. Countries such as the USA, UK, and some European countries specify full-scale façade testing but then permit exemptions for specific types of material based on small-scale fire testing. The United Arab Emirates has recently drafted and is applying regulations using full-scale façade testing combined with small-scale tests in response to a spate of fire incidents involving metal clad materials in 2011–2012. New Zealand primarily applies the cone calorimeter ISO 5660 for regulation of exterior walls. This appears to be the only country to do this. Some countries including Australia have no reaction to fire requirements except that the exterior walls must be non-combustible. However, in practice, combustible systems are applied as fire engineered performance-based designs (Alternative Solutions). In some countries, fire resistance tests are also required.

A range of different full-scale façade tests are in use around the world and have been reviewed. The geometry, fire source, specimen support details, severity of exposure, and acceptance criteria varies significantly for different tests. Existing research has identified that exposure to the exterior wall system is generally more severe for an internal post flashover fire with flames ejecting from windows than for an external fire source. For this reason, almost all of the full-scale façade fire tests simulate an internal post flashover fire. However, it is possible for the severity external fires at ground level on fuel loads such as back of house storage areas and large vehicle fires to equal or exceed internal post flashover fires. Although most full-scale façade tests simulate an internal post flashover fire, they may also set a suitable level of performance with regard to external fires.

Full-scale façade tests are currently the only method available for absolutely determining the fire performance of complete assemblies which can be influenced by factors which may not be adequately tested in small-scale tests. These factors include the severity of fire exposure, interaction of multiple layers of different types of materials, cavities, fire stopping, thermal expansion, fixings, and joints. However, full-scale tests are usually very expensive. Based on the present review, we note that:

- Dimensions and physical arrangement of facade tests vary. As an example, some large-scale tests involve external corner walls 8 m high (UK) or 5.7 m high (Germany and ISO) and 2.4 m and 1.3 m wide.
- There are significant differences in the source fire simulating a fire in the room of origin. Wood cribs, liquid pool fires, and gas burners are being used to generate maximum heat fluxes on the façade in the range of 20–90 kW/m². It needs to be investigated if these fires represent a sufficient and reasonable exposure to represent real-fire scenarios.
- Test durations, measurements, and acceptance criteria vary.
- The degree to which suitability of fixing systems and fire spread through joints, voids, and window assemblies of a multifunctional façade assembly are tested varies.
- Whilst large-scale facade tests do not measure key flammability properties of the individual elements of the facades, these tests do provide useful information for validation of fire spread modelling.

Intermediate-scale tests including ISO 13785 Part 2, the Vertical channel test and also a variety of room corner tests have been reviewed. These are less expensive, however, they may not correctly predict real-scale fire behaviour for all types of materials due to less severe fire exposures, less expanse of surface material to support fire growth and flame spread, and less incorporation of end use construction such as joints, fire stopping, and fixings. Except for the SBI test, intermediate-scale tests are currently not used for regulation, however, they are a cost-effective method for product development. The SBI test is currently typically applied to individual façade components rather than whole assemblies.

Small-scale tests applied for regulation of exterior wall materials around the world have been reviewed. Small-scale tests often are only applied to individual component materials and represent very specific fire exposure conditions. Small-scale tests can provide misleading results for materials which are complex composites or assemblies. This is particularly the case where a combustible core material may be covered by a non-combustible or low-combustible material or a highly reflective surface. There is currently no practical method of predicting real-scale fire performance from small-scale tests for the broad range of exterior wall systems in common use. Small-scale tests may provide acceptable benchmarks for individual material components. However, further validation against full-scale tests may be required to support this. Small-scale tests (in particular, the cone calorimeter) can also be useful for doing quality control tests on materials for systems already tested in full-scale or for determining key flammability properties for research and development of fire spread models. Small-scale tests, such as the cone calorimeter, should not be used to assess the performance of the whole façade assembly.

Development of a new full-scale test to simulate a specific fire scenario is not recommended at this stage. Instead further research to validate the existing full-scale and small-scale tests and also to develop a more affordable and dependable intermediate-scale test are recommended. A range of options for further test-based research for Phase II have been proposed. In summary, these are:

- Option 1: Existing full-scale façade test round robin—conduct tests on the same wall assembly applying the different large-scale tests currently operated by labs around the world. This would increase understanding of the relative performance of the different test method, provide a basis for accepting systems tested under different methods, and would provide full-scale data to support other research options suggested.
- Option 2: Development and validation of intermediate-scale façade test—This may possibly enable reliable regulation of materials using a less expensive test or at least enable less expensive testing for product development.
- Option 3: Validation of small-scale test regulatory requirements against large-scale tests. This would include collating any exiting small-scale and full-scale test data on a range of exterior wall systems that can suitably be applied to validate requirements. Indentify and carry out any further small-scale and full-scale testing that may be required to validate requirements. Examine test data to investigate any correlations and limitation of small-scale tests vs. large-scale performance and conclude on the suitability of existing regulatory requirements. Investigate more appropriate ways to test individual façade components which in combination with proper fire breaks would give a better assessment of the behaviour of a full-scale façade.

- Option 4: Investigation of vertical "U" channel on full-scale test—fire incidents indicate that very rapid fire spread may result for external vertical "U" shaped channels extending over a significant height of the building created by balconies and the like on the exterior. Modifying an existing full-scale test would enable investigation the impact this profile has on materials which pass in standard test geometry and if any increased requirements are needed for materials that are to be installed in this arrangement in end use. Assessment of this situation and the development of such a test ("U" shaped façade with side wing walls) may be assisted by the recent work by FireSERT and USTC (China) on facade flame heights with side walls [2].
- Option 5: Development of façade flame spread models—continued research on developing and validating flame spread models is required to move beyond current limitations.

An alternative or parallel performance-based approach, which can also be used for risk analysis, is also proposed:

1. Assess and measure key flammability properties of the combustible facade components in small- (cone calorimeter) and intermediate-scale experiments (SBI). Based on these tests and analysis, classify, for example, the materials according to European regulations for construction products. Then for regulation, Euro class B or better may be accepted for individual components.
2. Determine size of fire for the specific enclosure in the built environment based on recent research work.
3. Reproduce this fire size using a gas burner in a test similar, for example, to one proposed and developed in Japan as Option 2.
4. Measure and/or model the heat fluxes of facade flames on an inert façade in the selected test.
5. Test the real facade assembly and use the results to assist in establishing regulations.
6. If load bearing facade, perform also a fire resistance test with conditions reproducing the heat fluxes in Part 3.

Highett, VIC, Australia                                                                               Nathan White
Jordanstown, Ireland                                                                       Michael Delichatsios

# Acknowledgments

The Research Foundation expresses gratitude to the report authors Nathan White, CSIRO, and Michael Delichatsios, University of Ulster. The Research Foundation appreciates the guidance provided by the Project Technical Panelists, the funding provided by the Property Insurance Research Group (PIRG), and all others that contributed to this research effort.

**Project Technical Panel**

Liza Bowles, Newport Partners LLC
Dave Collins, Preview Group
Doug Evans, Clark County Building
Todd Gritch, HKS Architects
Jeff Harrington, Harrington Group
Gavin Horn, University of Illinois
Robert Jansson, SP
Richard Keleher, Thompson Litchner
Birgitte Messerschmidt, Rockwool
Faimeen Shah, Vortex Fire Engineering Consultancy
Robert Solomon, NFPA
Tracy Vecchiarelli, NFPA

**Project Sponsors**

CNA Insurance
FM Global
Liberty Mutual
Tokio Marine Management, Inc.
Travelers Insurance
XL Group
Zurich NA

## About the Fire Protection Research Foundation

The Fire Protection Research Foundation plans, manages, and communicates research on a broad range of fire safety issues in collaboration with scientists and laboratories around the world. The Foundation is an affiliate of NFPA.

## About the National Fire Protection Association (NFPA)

NFPA is a worldwide leader in fire, electrical, building, and life safety. The mission of the international nonprofit organization founded in 1896 is to reduce the worldwide burden of fire and other hazards on the quality of life by providing and advocating consensus codes and standards, research, training, and education. NFPA develops more than 300 codes and standards to minimize the possibility and effects of fire and other hazards. All NFPA codes and standards can be viewed at no cost at www.nfpa.org/freeaccess.

# About the Book

This report has been prepared by CSIRO (Commonwealth Scientific Research Organisation, Australia) and FireSERT (University of Ulster) for the Fire Protection Research Foundation (FPRF) as the deliverable for the project *Fire Hazards of Exterior Wall Assemblies Containing Combustible Components*.

CSIRO advises that the information contained in this publication comprises general statements based on scientific research. The reader is advised and needs to be aware that such information may be incomplete or unable to be used in any specific situation. No reliance or actions must therefore be made on that information without seeking prior expert professional, scientific, and technical advice. To the extent permitted by law, CSIRO (including its employees and consultants) excludes all liability to any person for any consequences, including but not limited to all losses, damages, costs, expenses, and any other compensation, arising directly or indirectly from using this publication (in part or in whole) and any information or material contained in it.

# Contents

# Chapter 1
# Introduction

## 1.1 Background

Many combustible materials are used today in commercial wall assemblies to improve energy performance, reduce water and air infiltration, and allow for aesthetic design flexibility. There have been a number of documented fire incidents involving combustible exterior walls but a better understanding is needed of the specific scenarios leading to these incidents to inform current test methods and potential mitigating strategies.

The Fire Protection Research Foundation funded a research project on 'fire hazards of exterior wall assemblies containing combustible composites'. The background to the project stated:

*This project will review available fire statistics, fire incidents, and literature and test methods relating to combustible external wall assemblies. Where possible the review will focus on each of the below identified types of materials separately. It is understood that a review of EIFS is of particular interest to the project sponsors and this will be provided as well as a review of the other types of assemblies identified.*

- *Exterior Insulation Finish Systems (EIFS) or synthetic stucco*
- *Metal composite material cladding (MCM)*
- *High-pressure laminates*
- *Structural Insulation Panel Systems (SIPS) and insulated sandwich panel systems*
- *Weather-resistive barriers (WRB).*
- *External timber panelling and facades including cross laminated timber (CLT) may also be increasing in use to increase renewable composition of buildings.*

*Other important issues that will be examined in the review are the types of buildings where these materials are used and how this affects fire incidents and hazards.*

© Fire Protection Research Foundation 2015
N. White, M. Delichatsios, *Fire Hazards of Exterior Wall Assemblies Containing Combustible Components*, SpringerBriefs in Fire,
DOI 10.1007/978-1-4939-2898-9_1

## 1.2   Objective

To develop the technical basis for evaluation, testing and fire mitigation strategies for exterior fires exposing exterior wall systems with combustible components.

## 1.3   Scope of Work

The Fire Protection Research Foundation separated the research into two phases which was described as follows:

- *Phase I—Review of fire incidents and statistics, literature and relevant test methods. Identification of selected fire scenarios and test methods for Phase II*
- *Phase II—Experiments/tests to evaluate performance of exterior walls with combustible materials*

    *Phase 1 is broken into the following tasks:*

*Task (a)  With the assistance of NFPA's Fire Analysis Division, conduct a review of the national fire incident reporting system database as well as other databases and compile information on typical exterior fire scenarios which involve the exterior wall.*

*Task (b)  Conduct an informal survey of fire departments and the fire service literature to identify fire incidents involving exterior wall systems with combustible materials to gather further case study information.*

*Task (c)  Compile and evaluate relevant test methods and listing criteria and other approval/regulatory requirements for these systems.*

*Task (d)  Compile the information from Tasks (a) to (e) into an information bulletin on combustible exterior wall fire safety.*

*Task (e)  Using the results from (a) to (g), identify selected fire scenarios and testing approach for Phase II evaluation of the fire performance of exterior walls with combustible materials. These scenarios should reflect real world conditions and include the potential for evaluation of the effectiveness of external fire protection features.*

    The scope only covers Phase I.

## 1.4   Limitations

- The study excluded single family residential dwellings (houses).

# Chapter 2
# Combustible Exterior Wall Systems in Common Use

The following provides an overview of some of the most common types of exterior wall assemblies containing combustible materials and typical fixing methods. It is noted that the following assemblies can typically be installed as:

- External cladding or covering over a solid structural external wall; or
- A curtain wall system, which is a non-structural external covering spanning multiple floors which are typically supported by mounts at the edge of each floor. This typically results in a gap between the edge of each floor and the curtain wall.

## 2.1 Exterior Insulation Finish Systems (EIFS)

Exterior Insulation Finishing Systems (EIFS) are sometimes referred to as "synthetic stucco" or External Thermal Insulation Composite Systems (ETICS). EIFS are attached to the exterior wall substrate to improve thermal insulation, weather tightness or for aesthetics. EIFS may be applied to masonry and concrete walls or lightweight walls typically lined with a suitable substrate such as gypsum board or cement board. A wood substrate with a waterproof membrane is used in some cases. The EIFS is typically attached with an adhesive (cementitious or acrylic based) or mechanical fasteners.

EIFS consists of a number of layers that are installed in the following order. The most basic EIFS consists of three layers:

- A layer of insulation, usually foamed polymer. Most EIFS use expanded polystyrene (EPS), however other types of foamed polymers are sometimes used including phenolic, polyisocyanurate (PIR) or polyurethane. The insulation layer is typically 1/4"–4" thick however thicker layers of 8" or higher are becoming more

© Fire Protection Research Foundation 2015
N. White, M. Delichatsios, *Fire Hazards of Exterior Wall Assemblies Containing Combustible Components*, SpringerBriefs in Fire,
DOI 10.1007/978-1-4939-2898-9_2

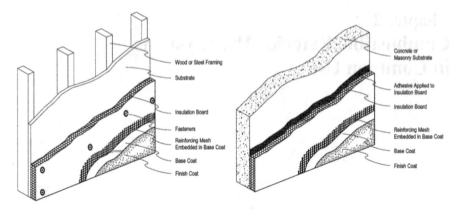

**Fig. 2.1** Typical EIFS applied to light weight framed construction (*left*) and Masonry construction (*right*) [3]

common, particularly in Europe. Non combustible insulation materials such as stone wool can also be used.
- A reinforcing mesh layer. Typically fibre glass reinforcing mesh.
- A base coat and top coat finish. Typically applied with a trowel or sometimes sprayed. The coating material is typically a cement based polymer modified render that is resistant to ignition and combustion.

In some modern EIFS installations a drainage or water management system is included. In this case a water resistive barrier (membrane) is installed over the substrate and small drainage cavities are created between the membrane and the foam when installed (Fig. 2.1).

Systems are available which use EPS as the formwork for poured concrete walls. The remaining EPS formwork is then rendered over on the external surface and covered with plasterboard on the internal surface.

## 2.2   Metal Composite Material Cladding

Metal composite claddings are typically thin section panels also known as Aluminium Composite Material (ACM). Typically they consist of two 0.5 mm thick aluminium sheets with a core material sandwiched between. The core material thickness typically ranges from 2 to 5 mm thick. The core material is typically either polyethylene or a mineral filled core which typically consists of polyethylene with a percentage of mineral filler. A high ratio of mineral filling provides significant improvement in fire performance. The surface is typically coated with a fluorocarbon surface coating in a range of different colours. These panels are significantly less expensive than solid metal panels at a thickness required to achieve the same flexural stiffness (Fig. 2.2).

**Fig. 2.2** Typical metal
composite claddings [4]

Metal composite panels are typically installed to exterior walls on steel channels or battens/top hats. This can create an air gap (typically about 40 mm) between the wall surface and the cladding. The panels are typically fastened to the steel battens by either of the following two methods.

- Flat stick method—panels adhered to steel battens using double sided adhesive tape
- Cassette mount method—the edges of the panels are folded at right angles and are rivet or screw fixed to aluminium or steel channels or clips which are in turn screw fastened to the exterior wall.

Sealant is normally applied to the gaps between panels. The above type of installation typically forms a ventilated façade/rain screen with an air gap separating the metal composite panel from the supporting wall behind.

## 2.3   High-Pressure Laminates

Exterior grade High Pressure Laminate (HPL) Panels are typically layers of phenolic resin impregnated cellulose fibres (typically up to 70 % cellulosic fibre content) with one or more decorative surface layers which are manufactured by pressing at high temperature and pressures typically >1000 lb per square inch (70 kg/cm$^2$). This high pressure and temperature is required for the thermosetting poly-condensation process of the resin used. A wide range of colours, patterns and surface textures for the decorative surface layer are possible. The resulting panel dense with a good

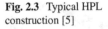

**Fig. 2.3** Typical HPL
construction [5]

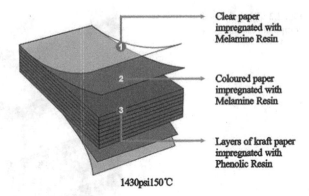

Clear paper
impregnated with
Melamine Resin

Coloured paper
impregnated with
Melamine Resin

Layers of kraft paper
impregnated with
Phenolic Resin

1430psi150°C

strength to weight ratio and is extremely weather resistant. HPL panels are typically available in thicknesses ranging from 3 to 14 mm. HPL panels are typically applied as ventilated facades/rain screens, balcony panels and sun louvers.

HPL panels are typically installed over the existing wall surface using metal channels (battens or top hats) to separate the panel from the supporting wall. The panels at fixed to the metal channels either by exposed screws or rivets or on thicker panels (8 mm or thicker) concealed screwing of mounting clips to the inside of the panel is possible (Fig. 2.3).

## 2.4  Structural Insulation Panel Systems (SIPS)/Insulated Sandwich Panel Systems

Structural insulated panels (SIPs) or insulated sandwich panels are typically used for walls but can also be used for ceilings, floors, and roofs. Sandwich panels may be divided into two major groups;

1. Metal skinned sandwich panels—These are composed of thin steel skins (0.4–0.6 mm thick) adhered to both sides of an insulating core which may typically be Expanded Polystyrene (EPS), Poly-Isocyanuarate (PIR), Phenolic, EPS beads in a phenolic matrix or mineral fibre. The total thickness of panels range from 50 to 200 mm. Metal skinned sandwich panels are typically used in food processing and storage facilities due to their high thermal resistance, durability and inert and easily cleaned surface. The metal skin provides a degree of resistance to ignition of the core materials however this is significantly affected by the penetrations, jointing and fastening systems used. In the event of large fires such panels can pose a hazard of collapse as the panels lose structural stiffness if the core materials melt or soften [6].

**Fig. 2.4** Typical sandwich panels with EPS core (*left*), Rockwool core (*centre*) and compressed straw with cardboard skin (*right*)

2. Non-Metal skinned sandwich panels—These are typically composed of skins such as plywood, oriented strand board, paper or cardboard based products, Gypsum or cement board. These are adhered to an insulated core of the same materials as listed above. Alternatively compressed straw is sometimes used as the core material. These types of sandwich panels are typically used in low rise residential, education and public assembly type buildings. They are sometimes used for acoustic properties as well as thermal properties.

Sandwich panels have been applied as a cladding to the exterior surface for building types other than low risk food processing and warehouses. For example schools, hospitals, prisons, retail outlets and other public buildings have made use of this material (Fig. 2.4) [7].

## 2.5   Rain Screen Cladding (RSC) or Ventilated Facades

Rain screen cladding, sometimes referred to as a ventilated façade, is a type of façade construction which typically includes the following elements

- The external wall/substrate—this may be solid masonry or concrete construction or a light weight framed wall lined with an exterior grade sheeting product such as gypsum or cement board or timber board products with a water proof membrane
- Insulation fixed to the exterior of the substrate—Typically panels of mineral fibre based insulation or foamed phenolic, polyisocyanurate (PIR), expanded polystyrene (EPS), or polyurethane (PU) may be adhered or mechanically fastened to the substrate. In some cases a spray based insulation may be applied
- Ventilation cavity and supporting brackets—a ventilation cavity of at least 25 mm typically exists between the insulation and the rain screen external cladding. The cladding is supported by aluminium or steel brackets which bridge across the air gap.
- Rain screen cladding panel—A wide range of materials are typically used including metal composite cladding, high pressure laminates, timber products, metal

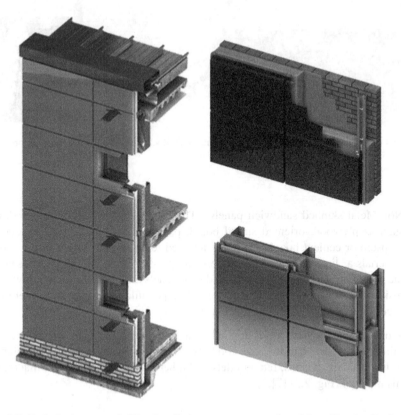

**Fig. 2.5** Typical rain screen cladding installation arrangements—from Linear Facades Catalogue [8]

sheeting, ceramic tiles, and cement board products. The cladding may include gaps between edges of panels and usually includes significant openings at the top and bottom of the wall to promote ventilation and drainage though the cavity.

Rain screen cladding can be applied during primary construction or as refurbishment to existing construction (Fig. 2.5).

Rain screen cladding systems are usually installed due to the following possible benefits;

- Improved protection against moisture ingress into buildings
- Improved thermal performance through solar shading, increased insulation and reduced thermal bridging.

## 2.6   Weather-Resistive Barriers and Combustible Wall Cavity Insulation

Weather resistive barriers are typically installed within the wall cavity to control air and moisture transmission and in some cases provide insulation to radiant or conducted heat transfer. Weather resistive barriers come in the following forms (Fig. 2.6).

- Mechanically attached membrane known as sarking or building wrap. Typically this is made out of woven bonded polyethylene fibre.
- Self adhering membranes.
- Fluid/paint applied membranes which include polymeric and asphaltic based materials
- Spray applied polymeric foams such as polyurethane which also provide insulation
- Board type barriers which includes plywood (typically up to 12 mm thick) or foamed plastic boards such as EPS or phenolic up to 25 mm thick (sometimes with a foil facing)
- Cellular insulation wraps which typically are made of polyethylene and have an air bubble structure much like bubble wrap. These often come with a reflective foil facing. They are typically 4–10 mm thick

**Fig. 2.6** Typical weather resistive barriers including sarking (*top left*), air cell insulation (*bottom left*) and foil faced EPS board (*right*)

## 2.7   External Timber Panelling and Facades

Motivation for increased use of timber based materials exists due to increase the renewable composition of buildings. In addition to traditional timber cladding and building materials, cross laminated timber (CLT) is also increasing in use. CLT is constructed of layers of timber, known as lamellas glued and pressed together with the grain alternating at 90° angles for each layer. Individual thicknesses of layers start at 10 mm. Number of layers is typically 3, 5, or 7. Total thickness ranges from 57 to 400 mm. CLT is used similar to tilt up concrete panels for structural walls, roofs and floor slabs. Often the external walls will be covered by a façade/cladding for weather resistance however CLT may also be used as an exposed feature with appropriate protective coatings (Fig. 2.7).

**Fig. 2.7**  Typical CLT Panels (*left*), Forte 10 storey residential CLT building in Melbourne (*right*)

# Chapter 3
# Existing Research and Mechanisms of Fire Spread

The following section presents an overview of existing research literature on fire performance of exterior combustible wall assemblies. This section also identifies the key mechanisms of fire spread on combustible exterior wall assemblies.

## 3.1 Existing Research

It is a difficult task to provide due credit to all previous work related fire performance of exterior combustible walls. A large amount of literature has been generated in this area over past decades. For this reason, we have included two lists of references:

- Appendix D is a list of related research literature. It is not practical to summarise all of this research, however this list is provided for further reading.
- The References section of this report lists references that are directly referenced.

In 2000, the Fire Code Reform Centre funded a research report on fire performance of exterior claddings [1]. The scope of the FCRC report is very similar to the scope of this book. It presents a review of façade ignition and fire spread scenarios, key fire incidents, previous research, regulations and test methods and performance of typical materials in various test methods.

The FCRC report recommended

- Collection of fire incident data should be modified to identify and capture details of external vertical fire spread.
- The intermediate scale ASTM Vertical channel test should be developed and adopted by Australian and New Zealand building codes as a cost effective means of evaluating exterior wall systems. Full scale façade tests were considered to be too costly.

© Fire Protection Research Foundation 2015
N. White, M. Delichatsios, *Fire Hazards of Exterior Wall Assemblies Containing Combustible Components*, SpringerBriefs in Fire, DOI 10.1007/978-1-4939-2898-9_3

The above recommendations were not adopted in Australia. Some regulations and test methods worldwide have been revised since 2000.

The review of previous research presented by the FCRC report is very thorough and provides an excellent overview of the research in different countries on façade fire behaviour and test development as well as performance of insulated sandwich panels, curtain wall systems, EIFS, apron and spandrel protection and façade fire calculation tools and modelling up to the year 2000. As this document is still publically available and can be requested from ncc@abcb.gov.au this current report will not repeat this review but will instead focus on key developments and research since the year 2000.

A 2011 report by Exova Warringtonfire Australia [9] provides a general guideline for fire safety engineering design of combustible facades. It addresses three façade fire exposure scenarios of fire spread from an external fire source, fire and smoke spread from an internal compartment fire and fire spread to adjacent structures. The report outlines methods of calculating the resulting radiant heat or direct flame exposure for given scenarios. It also reviews existing data on critical radiant heat flux for piloted ignition and resistance to fire spread for typical materials. It recommends that fire engineers should also use test data from existing test protocols including radiant heat exposure tests such as AS1530.4 Appendix B7 ($3 \times 3$ m fire resistance furnace enclosed with steel sheet to act as radiant heat source) [10], fire resistance tests, intermediate scale façade fire spread tests such as ISO-13785-1 [11] or full scale faced tests such as ISO 13785-2 [12]. The Exova Warringtonfire report also presents results of experiments using the ISO-13785-2 apparatus with non-combustible walls to compare radiant heat and flame height measurements to prediction calculations based on Laws Method [13–15], and also investigate the reduction in fire exposure due to installing a 0.6 m non-combustible horizontal projection above the window opening.

The 2012 Fire Protection Research Foundation report on Fire Safety Challenges of Green Buildings [16] identifies green building design elements that increase fire safety hazards and best practices for hazard risk mitigation. This includes external combustible wall assemblies such as double skinned façades, EIFS and SIPS. This report provides references to some fire incidents involving combustible external wall assemblies.

The 2013 BRE report "BR 135" [17] on fire performance of external thermal insulation provides a good overview of UK regulations, mechanisms of external fire spread, fire performance design principles for external cladding systems and a summary of BS 8418-1 and BS 8414-2 test methods and performance/classification criteria. Based on extensive BRE test programs, it is concluded that fire spread on the wall system should be contained to the floor immediately above the floor of fire origin. The importance of fire barriers at each level, both set into the insulation and across ventilated cavities (whilst still enabling ventilation air movement) his highlighted. This document provides detailed guidance on suitable design and installation of combustible external insulation facades.

A 2013 report by VTT [18] investigates the effect of the use of EPS based EIFS on the fire safety of multistorey residential buildings. This investigation reviews statistical data from Finland and Sweden and uses probabilistic event tree based risk analysis to assess the risk of fire spread between floors (see Sect. 4.2.1). This report also summarises experiments and CFD modelling to determine heat flux exposure to external walls for flames emerging from typical dwelling room flashover fires. It was concluded that in these scenarios the heat flux to a window on the level above may initially spike to as high as $120 \, kW/m^2$ when the fire compartment window first breaks. After this initial spike the peak heat flux over the duration of the fire may be a maximum of $80 \, kW/m^2$. It was also concluded that the heat flux to the window two levels above the fire compartment are typically about one third of the heat flux one level above. For external fire sources such as car fires or waste bins a heat flux exposure of $30$–$40 \, kW/m^2$ at a distance of 1 m was considered reasonable and on this basis the room flashover fire was considered to provide the most sever exposure to the façade.

A series of relevant papers are available as the Proceedings of the 1st International Seminar for Fire Safety of Facades [19].

Research by FM Global has examined expected heat exposures to EIFS wall systems from external storage for commercial/industrial buildings [20]. Existing full-scale façade tests for EIFS which simulate flames issuing from an open window where found to be less than expected for this exterior fire scenario and the FM approvals 50-ft Corner test allowed practical procedures for the categorisation of EIFS systems.

## 3.2 Mechanisms of Fire Spread

Based on review of reported fire incidents and existing research the following key types of initiating fire events and types of fire spread after the initiating event have been identified:

Key Initiating events

- Interior fire (pre flashover or post flashover) spreading to external wall system via external openings such as windows.
- Interior fire (pre flashover or post flashover) spreading to external wall system via internal openings including cavities and concealed spaces.
- Exterior fire directly adjacent external wall system igniting the wall due to radiant heat and/or flame impingement
- Exterior fire spatially separated from external wall system resulting in radiant heat only (fire in adjacent building for example)

Key mechanisms of fire spread after initiating event

- Fire spread to the interior of level above via openings such as windows causing secondary interior fires on levels above resulting in level to level fire spread
- Flame spread over the external surface of the wall
- Flame spread within an interval vertical cavity/air gap

- Heat flux impacts cause degradation/separation of non-combustible external skin resulting on flame spread on internal core
- Secondary external fires to lower (ground) levels arising from falling burning debris.

The key initiating fire may be simply summarised as one of two possible types of fires:

- Fires external to the building (other burning buildings, external ground fires) or
- Fires internal to the building originating in a floor that have resulted in breaking the windows and ejecting flames on the façade as in Fig. 3.1 or fire spread to wall cavities.

The existing research has identified that exposure to the exterior wall system is generally more severe for an internal post flashover fire with flames ejecting from windows than for an external fire source. For this reason, most full scale façade fire tests simulate an internal post flashover fire. However it is possible for the severity external fires at ground level on fuel loads such as back of house storage areas and large vehicle fires to equal or exceed internal post flashover fires. The impact of exterior fire sources can be even more severe if they occur hard against re-entrant exterior wall corners. Although most full-scale façade tests simulate an internal post

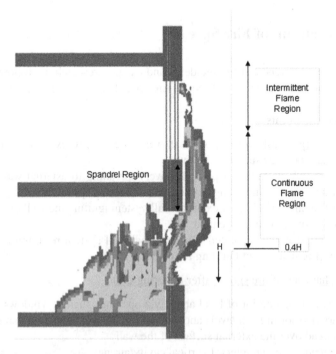

**Fig. 3.1** Enclosure fire and floor to floor fire spread

flashover fire, these tests may also set a suitable level of performance with regards to a limited external fire severity.

In order to design for mitigation of the facade fire spread hazard, one should be able to determine the size of the fire in the enclosure, the ejected facade flame properties (flame height and heat fluxes) and their impact on facade assemblies and building materials. These issues have been addressed to various degrees of completeness experimentally, analytically and numerically by various investigators as one can see for example in the proceedings of facade flame conference [19] as well as in references in Appendix A.

It follows from the previous brief discussion and the other sections of this report that the façade fire safety problem (and existing research) can be divided in four parts:

1. Specification of fire development and the heat flux distribution both inside the enclosure and from the façade flames originating from the fire in the enclosure. This requirement is prerequisite for the following parts.
2. Fire resistance of the façade assembly and façade—floor slab junction including structural failure for non-combustible and combustible façade assemblies
3. Fire spread on the external surface of the façade assembly if combustible due to the flames from the enclosure fire
4. Fire spread and propagation inside the façade insulation, if combustible, due to the enclosure fire.

We address these parts of the façade fire safety problem next.

Heat fluxes and flame heights from enclosure fires and façade flames (Part 1)

Recent reviews of enclosure and façade fires and flames [21, 22] address Part 1 of the façade fire safety problem as outlined above. In these references, it is shown that under ventilated fires generate the larger hazard regarding the heat flux impact and flame heights both inside and outside the enclosure. Also a methodology based on dynamic similarity developed by experiments is provided to calculate these heat fluxes having agreement with the work on large scale fires of Oleszkiewicz [23, 24]. This methodology is applied in detail in a Master thesis at FireSERT [25]. Figure 3.1 shows the schematic cross section of a typical enclosure plume fire from an opening.

Fire Resistance of the façade assembly for load and non-load bearing situations (Part 2)

Given the magnitude of heat flux impact from Part 1, it is possible to assess the structural fire resistance of the façade assembly and the façade-floor junction for non-combustible components by performing standard fire resistance tests in conjunction with structural analysis depending on the properties of the components if known (e.g., steel, concrete, masonry, plasterboard) and on the complexity of the assemblies.

If the façade assembly has combustible components (e.g. insulation, render etc), fire resistance assessment has still to be performed but now following evaluation of the fire spread hazards as required for Parts 3 and 4 which can create additional heat impact on the façade and façade–floor assemblies.

Fire spread on the external surface or inside the façade assembly for combustible components (Parts 3and 4)

To address this hazard rationally one would need to assess first the flammability of the combustible components of the façade assembly and then perform a large scale façade test (>4 m high) for the whole façade assembly designed to reproduce the actual heat fluxes and flame heights determined in Part 1 for a specific type of enclosures, fuel loads and openings.

One way to characterise the flammability of the combustible components (polymer, render, wood, sealants) is to determine their key flammability properties using the cone calorimeter (and possibly, TGA/FTIR/DSC). Although these small scale tests can be used to measure flammability of individual combustible components, they are not always capable of directly predicating full scale performance of composite systems with multiple layers, joints and possibly internal fire stop barriers. For this reason many countries apply full-scale façade tests for the purposes of regulation.

There are several examples of research attempting to either develop suitable intermediate scale tests or to apply and validate CFD models together with small scale test results to predict fire behaviour in a larger scale test such as SBI (Single Burning Item) [26]. The aim of this is to assess the hazard of the combustible components prior to performing a large scale (>4 m high) façade test.

# Chapter 4
# Fire Statistics

## 4.1 U.S. Fire Statistics 2007–2011

### 4.1.1 Methodology

A preliminary statistical analysis of building fires reported to U.S. municipal fire departments has been completed for fire incidents relating to exterior walls.

The 2007–2011 statistics in this analysis are projections based on the detailed coded information collected in Version 5.0 of the U.S. Fire Administration's (USFA's) National Fire Incident Reporting System (NFIRS 5.0) and the findings of the National Fire Protection Association's (NFPA's) annual survey of local or municipal fire department experience [27, 28].

The number of NFIRS code choices relating to exterior wall fires is very limited and does not capture information such as the type of exterior wall material (combustible or non-combustible), the extent of fire spread, or the mechanism of fire spread (external surface or within cavity).

Except for property use and incident type, fires with unknown or unreported data were allocated proportionally in all calculations. Casualty and loss projections can be heavily influenced by the inclusion or exclusion of one or more unusually serious fires. Property damage has not been adjusted for inflation. Fires, civilian deaths and injuries are rounded to the nearest one and direct property damage is rounded to the nearest hundred thousand dollars (US).

The following property type use codes were included:

- Public assembly (100–199)
- Educational (200–299)
- Health care, nursing homes, detention and correction (300–399)
- Residential, excluding unclassified (other residential) and one-or two-family homes (420–499). This includes hotels and motels, dormitories, residential board and care or assisted living, and rooming or boarding houses.

© Fire Protection Research Foundation 2015
N. White, M. Delichatsios, *Fire Hazards of Exterior Wall Assemblies Containing Combustible Components*, SpringerBriefs in Fire,
DOI 10.1007/978-1-4939-2898-9_4

- Mercantile (500–589)
- Office buildings, including banks, veterinary or research offices, and post offices (590–599)
- Laboratories and data centres (629, 635, and 639)
- Manufacturing or processing (700)
- Selected storage properties: Refrigerated warehouses, warehouses, other vehicle storage, general vehicle parking garages, and fire stations (839, 880, 882, 888, and 891)

Separate queries were performed for:

- Fires starting in or on the exterior wall surface area (area of origin code 76), and
- Fires that did not start on the exterior wall area but the item first ignited was an exterior sidewall covering, surface or finish, including eaves, (item first ignited code 12); and
- Fires which did not start in the exterior wall or area or with the ignition of exterior sidewall covering but fire spread beyond the object of origin (fire spread codes 2–5) and the item contributing most to fire spread was the exterior sidewall covering (item contributing to flame spread code 12).

Results were summed after unknown or missing data, including extent of fire spread for the last condition, were allocated. This summed result is taken to represent the total number of exterior wall fires.

Separate queries were performed for four above ground height groupings:

- 1–2 stories,
- 3–5 stories,
- 6–10 stories, and
- 11–100 stories.

Separate queries were performed for four categories of automatic extinguishing system (AES) presence:

- Present [code 1],
- Partial system present [code 2],
- NFPA adjustment indicating AES presence but the reason for failure was the AES was not in the fire area [converted to code 8], and
- None present [code N,]

Review of this data must consider the context which is driven in a major way by the regulation in the USA. The USA does allow combustible facade materials and relies on several building codes and insurance requirements referring to a range of test methods and deemed to comply requirements. See Sect. 6.1.4 for further information.

## 4.1.2 Results

Table 4.1 shows the total number of "structure fires" in the selected property types overall, regardless of the area of origin or item first ignited. This includes fires with confined fire incident types (incident type 113–118), including cooking fires confined to the vessel of origin, confined chimney or flue fires, confined incinerator fires, confined compactor fires, confined fuel burner or boiler fires, and trash or rubbish fires inside a structure with no damage to the structure or its contents.

Table 4.2 shows the total shows the building fires (incident type 111) in selected properties that began on, at or with an exterior wall, by property use. These exclude the confined fire incident types listed above. Fires involving structures other than buildings (incident type 112), and fires in mobile property or portable buildings used as a fixed structure (incident type 120–123) were also excluded.

For all building types, Exterior wall fires accounted for 3 % of all structure fires. Exterior wall fires also accounted for 3 % of civilian deaths and injuries and 8 % of property damage. The highest number exterior wall fires occurred in residential buildings, and were 2 % of the total residential structure fires. However, the percentage of residential structure exterior wall fires was lower than the percentage of selected storage properties, public assembly, office buildings, and mercantile properties, with exterior wall fires being 10 % of storage occupancy structure fires.

For exterior wall fires in the selected occupancies

- 42 % were fires starting on the exterior wall surface,
- 32 % were fires where the area of origin was not exterior wall, but item first ignited was exterior sidewall covering, and
- 26 % were fires where area of origin or item first ignited were not an exterior wall but the item contributing most to fire spread was an exterior wall.

**Table 4.1** Total structure fires in selected properties

| Property use | Fires | Civilian deaths | Civilian injuries | Property damage (US$ millions) | Portion of total fires (%) |
|---|---|---|---|---|---|
| Public assembly | 15,374 | 6 | 172 | 446.2 | 9 |
| Educational | 6012 | 0 | 90 | 105.1 | 3 |
| Institutional | 7153 | 6 | 182 | 59.6 | 4 |
| Residential | 121,651 | 485 | 4592 | 1548.8 | 68 |
| Mercantile | 15,198 | 20 | 287 | 724.8 | 9 |
| Office building | 3538 | 4 | 40 | 112.1 | 2 |
| Laboratory and Data centre | 234 | 0 | 10 | 22.5 | 0 |
| Manufacturing or processing | 5742 | 8 | 176 | 593.2 | 3 |
| Selected storage occupancies | 2930 | 8 | 45 | 230.7 | 2 |
| Total | 177,833 | 537 | 5595 | 3842.9 | 100 |

**Table 4.2** Exterior wall fires—Building fires in selected properties in which the area of origin, item first ignited or item contributing most to flame spread was an exterior wall

| Property use | Fires | Civilian deaths | Civilian injuries | Property damage (US$ millions) | Portion of total structure fires (%) |
|---|---|---|---|---|---|
| Public assembly | 706 | 0 | 6 | 30.8 | 5 |
| Educational | 127 | 0 | 0 | 2.8 | 2 |
| Institutional | 94 | 0 | 0 | 4.6 | 1 |
| Residential | 2889 | 18 | 133 | 197.2 | 2 |
| Mercantile | 891 | 0 | 5 | 31.1 | 6 |
| Office building | 210 | 0 | 3 | 7.6 | 6 |
| Laboratory and Data centre | 5 | 0 | 0 | 1.5 | 2 |
| Manufacturing or processing | 120 | 0 | 1 | 6.3 | 2 |
| Selected storage occupancies, | 303 | 0 | 0 | 13.1 | 10 |
| Total | 5346 | 18 | 148 | 295.0 | 3 |

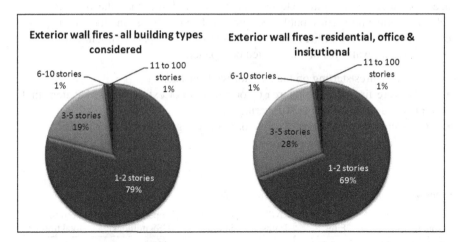

**Fig. 4.1** Percentage of exterior wall fires by building height

Inclusion of the exterior wall as the area of origin or item first ignited may be capturing scenarios such as fires in external fuel loads located against external walls or exposure of external walls to fires from adjacent buildings where the fire spreads to the interior of the building but the external (combustible or non-combustible) wall does not play a significant role in the fire spread.

The percentage of exterior wall fires within buildings of different height categories is shown in Fig. 4.1. This indicates that the vast majority of exterior wall fires occur within low rise (five stories or less) buildings. This may be due to two reasons:

• The majority of the building stock is low rise.
• Sprinklers are more likely to be installed in high rise buildings and reduce the risk of internal fires spreading via openings to the external facade.

As a sensitivity study, the percentage of exterior fires by building height has been plotted for only residential, office and institutional type buildings as these are expected to have a larger proportion of high rise building stock compared with other building types such as storage, manufacturing, mercantile and educational. A slightly increased percentage of exterior wall fires occur in three to five stories buildings compared to the other building types.

Figure 4.2 shows the percentage of exterior wall fires by presence of automatic extinguishing system within the different building height categories. Figure 4.2 indicates that the majority of exterior wall fires occur in buildings with no automatic suppression system or no automatic suppression system installed in the fire area. Two points need to be considered when examining this data. The NFIRS data element "presence of automatic extinguishing system" is intended to document "the existence of an AES within the AES's designed range of a fire. NFPA added the

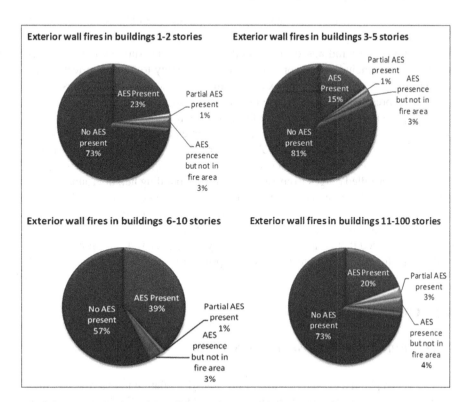

**Fig. 4.2**  Percentage of exterior wall fires by presence of automatic extinguishing system

category "present, but not in fire area, when an AES was coded as present, but the reason for a failure to operate was "Fire not in area protected."

Typical thresholds above which sprinkler systems are required in the *International Building Code* (IBC), 2012 Edition [29], include:

- Mercantile: Over 12,000 ft$^2$ (1115 m$^2$) in one fire area, or over 24,000 ft$^2$ (2230 m$^2$) in combined fire area on all floors, or more than three stories in height
- High-Rise: All buildings over 75 ft (22.86) m in height. However sprinklers are also required for all buildings with a floor level having an occupant load of 30 or more that is located over 55 ft (16.8 m) in height (ICC 903.2.11.3)
- Residential Apartments: Sprinkler protection required for all new residential apartment buildings

Typical thresholds above which sprinkler systems are required in NFPA 5000, *Building Construction and Safety Code*, 2012 Edition [30] include:

- Mercantile: Over 12,000 ft$^2$ (1115 m$^2$) in gross fire area or three or more stories in height
- High-Rise: All buildings over 75 ft (22.9 m) in height
- Residential Apartments: Sprinkler protection required for all new residential apartment buildings

Although it is expected that the majority of high-rise buildings (six stories or more) would have at least internal sprinkler systems, the majority of exterior wall fires for high rise buildings occur in buildings where no suppression system is installed. It is concluded that sprinkler systems are likely to have an effect on the risk interior fires spreading to the external wall to become exterior wall fires.

The data presented in Fig. 4.2 provides no information regarding failure of automatic suppression systems where installed. However, previous NFPA reports address sprinkler effectiveness in general. The data does not enable analysis of the effectiveness of internal sprinklers vs. external facade sprinklers in preventing exterior wall fire spread. The percentage of exterior wall fires occurring in buildings with sprinkler systems installed ranges from 15 to 39 % for the building height groups considered. This indicates that whilst sprinklers may have some positive influence, a significant portion of external wall fires still occur in sprinkler protected buildings, which may be due to both external fire sources or failure of sprinklers. On this basis it is recommended that controls on flammability of exterior wall assemblies should be the same for sprinkler protected and non-sprinkler protected buildings.

## 4.2   Other Fire Statistics

### 4.2.1   Finland and Sweden Fire Statistics

VTT has carried out a detailed review of fire statistics relating to exterior wall fires for multistorey residential buildings in Sweden and Finland [18]. The data reviewed was from 2004 to 2012 for Finland (average of 508 multistorey residential fires reported per year) and 2004–2011 for Sweden (average of 2739 multistorey residential fires reported per year). The regulatory requirements of these two countries are discussed in Sect. 6.1.3.

In summary, from this data it was concluded that:

- Approximately 88–92 % of ignitions are internal
- Approximately 2–3 % of ignitions are external ignitions other than balcony fires
- Approximately 5–6 % of ignitions are external ignitions on a balcony
- Fire spread outside the compartment of fire origin in 3 % of multistorey residential building fires
- Fires were suppressed or self extinguished without fire brigade intervention in about 15–30 % of fires.
- The probability of fire spread to between compartments via windows was estimated to be approximately 0.7–2 % from the statistical data
- The data relating to involvement of EPS insulation in fires was very limited. Only a few incidents were found that clearly involved EPS and these were mostly related to hot works, renovation or construction.

The VTT study also carried out a probabilistic event tree based risk assessment investigating and comparing the probability of fire spread between compartments via windows for EPS ETICs and low combustibility facades. This was calculated to be 2.3 % for EPS ETICS and 1.9 % for low combustibility facades.

It is noted that limitations relating to the use of small scale testing and reliance on FDS [31–33] fire modelling by this study may have resulted in errors in the prediction of the relative risk of fire spread on EPS ETICS vs. low combustibility facades.

It is also noted that the risk assessment was based on EPS ETICS installed with suitable rendering and mineral fibre fire barriers. The risk of non-compliant construction techniques as discussed in Sect. 5.1.2 was not evaluated.

## 4.2.2   New South Wales Fire Brigade Statistics, Australia

The Australian Incident Reporting System (AIRS) is an Australian national database framework for incidents reported to emergency services. Unfortunately not all Australian fire brigades actively report to AIRS and it is currently not well maintained or easy to retrieve data from.

New South Wales Fire Brigade (NSWFB) is one of the largest fire brigades in Australia. NSWFB publish annual fire statistics which represent a selection of the NSWFB AIRS data. The only information relating to exterior wall fires is the "area of fire origin" as shown in Table 4.3.

This indicates fires starting in wall assembly/concealed wall space are 0.5 % of the total fires and fires starting on exterior wall surfaces are 1.3 % of total fires. NSWFB statistics provide no information relating to the number of fires where the main area or fire spread was the exterior wall assembly or the types of exterior wall

**Table 4.3** NSWFB building fire statistics for area of fire origin [34]

| Area of fire origin | Year | | | |
|---|---|---|---|---|
| | 2003/2004 | 2004/2005 | 2005/2006 | 2006/2007 |
| Wall assembly, concealed wall space | 34 | 31 | 32 | 32 |
| Exterior wall surface | 82 | 77 | 95 | 69 |
| Total building fires (all areas of fire origin) | 6388 | 6165 | 6566 | 6257 |

**Table 4.4** NZFS building fire statistics for area of fire origin [35]

| Area of fire origin | Year | | | | |
|---|---|---|---|---|---|
| | 2009/2010 | 2008/2009 | 2007/2008 | 2006/2007 | 2005/2006 |
| Wall assembly, concealed wall space | 72 | 111 | 121 | 99 | 98 |
| Exterior wall surface | 224 | 321 | 355 | 307 | 290 |
| Total building fires (all areas of fire origin) | 4738 | 6361 | 6235 | 6269 | 6111 |

assemblies involved. The proportion of fire involving facades is lower than the other countries where statistics are available. This may be due to the non-combustibility requirements for buildings over 3–4 storeys. The risk correlates with the 1.9 % calculated by VTT for low combustibility facades.

### 4.2.3   New Zealand Fire Service Emergency Incident Statistics

The New Zealand Fire Service (NZFS) publish annual fire statistics in a similar format to NSWFB. Again, the only information relating to exterior wall fires is the "area of fire origin" as shown in Table 4.4.

This indicates that fires starting in wall assembly/concealed wall space are 1.7 % of the total fires and fires starting on exterior wall surfaces are 5.0 % of total fires.

The proportion of total fires involving facades is similar to that of the USA where both large and small scale test methods are applied in the control of combustible facades.

# Chapter 5
# Fire Incident Case-Studies

A literature review to identify reports for fire incident cases involving combustible exterior wall assemblies has been completed. This review has included

- Searches of fire engineering and fire science journals
- Internet and newspaper article searches
- Information on specific cases provided by technical review panel members for this project
- Fire brigades both in Australia (AFAC) and limited fire brigade contacts in the USA provided by the NFPA have been approached to provide fire incident reports, however none have been provided.

Literature for fire incident cases involving combustible exterior wall assemblies has been found to be limited. Generally only the most spectacular fire incidents are reported and most often the reports are in the form of newspaper articles with no specific information on materials, fire behaviour or mechanisms of fire spread. Very few documents presenting detailed investigations of the fire incidents have been found.

## 5.1 Fires Involving Exterior Insulation and Finish Systems

### 5.1.1 Miskolc, Hungary, 2009 [36]

On 15 August 2009, in Miskolc, Hungary, a fire started in a 6th floor residential kitchen resulting in vertical fire spread on the EIFS façade to the top of the 11 story building, resulting in 3 fatalities. The building was built in 1968 and was refurbished in 2007. The refurbishment included the installation of polystyrene based EIFS on the exterior walls. The fire also resulted in smoke spread through stair shafts and mechanical shafts (Fig. 5.1).

© Fire Protection Research Foundation 2015
N. White, M. Delichatsios, *Fire Hazards of Exterior Wall Assemblies Containing Combustible Components*, SpringerBriefs in Fire,
DOI 10.1007/978-1-4939-2898-9_5

**Fig. 5.1**  EIFS fire, Miskolc, Hungary, 15 August 2009 [36]

Post fire investigation indicated that the following issues contributed to the external fire spread.

- The system was not constructed in accordance with industry requirements
- Use of polystyrene insulation
- Inadequate sticking an fixing of lamina to polystyrene sheets
- No use of mineral wool insulation as fire propagation barriers, particularly around window reveals.

### 5.1.2  MGM MonteCarlo Hotel, Las Vegas, USA, 2008 [37–39]

The 32-story Monte Carlo Hotel and Casino was constructed in 1994 and 1995. The plan layout of the hotel was a centre tower from which three wings, each approximately 240 ft (73 m) long, extended. At the time of construction EIFS was installed as the exterior wall cladding. EIFS was installed in the flat areas of the building and on the decorative column pop-outs that extended from the 29th floor to the 32nd floor. Analysis indicated that these EIFS areas had a non-complying thickness of lamina (the exterior encapsulant) (Figs. 5.2 and 5.3).

Decorative non-EIFS architectural details constructed of EPS foam encapsulated in polyurethane resin were also installed on the exterior. These items included the horizontal band at the 29th floor, the horizontal band at the top of the 32nd floor, the railing at the top of the parapet wall and are believed to include the medallions between the windows on the 32nd floor.

**Fig. 5.2** MGM Montecarlo hotel fire, 2008 [39]

**Fig. 5.3** Decorative detail
constructed of EPS foam with
polyurethane resin
encapsulant [38]

The fire started before 11 am on 25 January 2008. Ignition of the exterior wall was attributed to welding on a catwalk on the roof parapet wall. The exterior cladding materials first ignited on the left side (as viewed from the exterior) of the central core area. The fire then progressed laterally. The adjacent materials on the right side of central core facade began to burn and the fire continued to propagate laterally over these decorative materials. The fire also moved to the left along the upper portion of the west tower and began to involve the cladding materials. Over time, the fire on the west tower moved laterally approximately 24 m. The fire did spread downwards, however not any further than the 29th floor. Heat from the exterior fire broke several windows however internal sprinklers halted any fire spread to the interior guest rooms. A total of 17 sprinklers activated.

Once the fire progressed away from the central core area, it appeared that the decorative band at the top of the 32nd floor, the medallions between the windows on the 32nd floor and the decorative band at the top of the wall were the primary mode of lateral flame propagation. Not only did these areas exhibit their own flame-spread, the resultant flames caused the flat area of the wall above to ignite.

The fire on the exterior facade was extinguished by the emergency responders at approximately 12.15 pm.

The fire was investigated by Clark County Nevada and Hughes associates [38]. Samples of materials were taken from the west wing of the building and the following was determined:

1. The horizontal band at the top of the exterior wall was over 5 ft (1.5 m) high and contained an expanded polystyrene (EPS) foam plastic core covered with a rigid non-EIFS coating.
2. The horizontal band above the uppermost guestrooms (32nd floor) was approximately 6 ft (1.8 m) high and contained up to 3 ft (0.9 m) thick EPS foam plastic at the upper portion covered with a polyurethane encapsulant.
3. The horizontal band at the 32nd floor was primarily hollow with a 13 mm thick outer shell comprised of fibreglass and a gypsum-plaster binder.
4. The horizontal band at the 29th floor was approximately 3 ft (0.9 m) high and contained 2 ft (0.6 m) thick EPS foam plastic at the top covered with a rigid non-EIFS encapsulant.
5. The decorative columns between the 29th and 32nd floors were approximately 2 ft 4 in. (0.7 m) wide and 6 in. (150 mm) thick with an EPS foam plastic core and an EIFS coating (thinner than required).
6. Each of the two base wall assemblies sampled contained 5/8 in. (16 mm) thick gypsum wall board covered with 1 in. of EPS foam. The primary difference between the two was that the EIFS coating from the upper sample was noticeably thinner but in both samples, the EIFS coating was thinner than required

The investigation resulted in the following conclusions

1. The primary contributor to the progression of the fire was the combination of materials in the decorative band at the top of the wall, the decorative band at the top of the 32nd floor (EPS with a polyurethane resin coating) and the undetermined materials in the medallions.
2. Flaming droplets and burning pieces of EPS and/or polyurethane caused ignition of the large decorative band at the 29th floor. This decorative band was composed of EPS and had anon-EIFS coating.
3. EIFS in the flat portion of the parapet wall was involved in the fire but was not the primary contributor to the lateral propagation of the fire, even though it appears to have a non-complying thickness of lamina. It did burn in the immediate area of fire exposure, as would be expected based on testing, but did not significantly propagate beyond the area of fire exposure caused by the burning of the decorative band at the top of the wall, the decorative band at the top of the 32nd floor and the medallions. As the fire progressed along these materials, it continued to involve the EIFS, but the EIFS was not the primary cause of the continued progression of the fire.

### 5.1.3   393 Kennedy St, Winnipeg, Manitoba, Canada, 1990 [1, 40, 41]

The building located at 393 Kennedy St was constructed in 1987 and is a 75 unit, 8 storey apartment building with a covered, open sided car park located at the ground floor with space for 54 cars. The building and car park was not sprinkler protected. The car park was not provided with fire detection. The car park was generally 2 h fire separated from the rest of the building (with the exception of open sides).

EIFS was applied to the exterior walls. The EIFS was applied to exterior grade gypsum sheathing on a steel stud frame or masonry wall. The foam insulation was typically 75 mm thick except in limit areas including the north façade where it was 140 mm thick. No Horizontal fire barriers (fire stopping) were included in the EIFS. The ceiling of the car park wall lined with a 65 mm thick rigid foam insulation covered by an aluminium soffit. Reports do not identify the type of foam insulation used in either the EIFS or car park ceiling. The north wall of the building faced an adjacent building located 3 m away (Figs. 5.4 and 5.5).

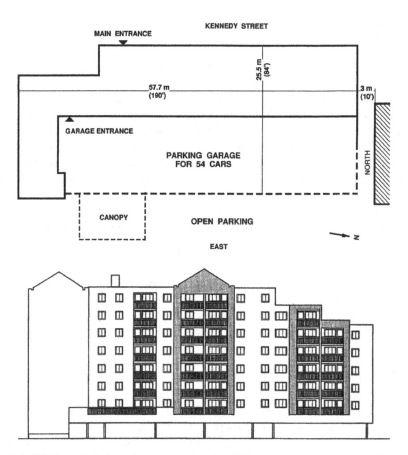

**Fig. 5.4**   393 Kennedy St floor plan and east elevation [40]

**Fig. 5.5**   393 Kennedy St East and North facades (north to right) post fire [40]

At 5.00 am on 10 January 1990, a fire started in the ground floor car park and very quickly spread to involve all 25 cars parked there at the time. The rapid spread of fire through the car park is attributed at least in part to the foamed plastic insulation on the ceilings. Flames issuing from the car park openings reached the third story (neglecting contribution from EIFS). The EIFS on the exterior walls was ignited and resulted in fire spread to the top of the fourth storey except for a narrow strip on the east facade where fire spread to the top of the seventh storey and on the north façade where the fire spread to the top of the building. It is noted that the EIFS were installed prior to the introduction of controls on EIFS in the Canadian building code and the EIFS installation is not expected to meet current code requirements. The enhanced fire spread on the north wall is believed to be due to the following factors;

- A south wind tended to drive flames across the car park to the opening in the north wall
- The close proximity of the adjacent building may have resulted in re-radiation and chimney effect
- The thicker foam insulation layer on this wall.

### 5.1.4   Dijon, France, 2010 [42]

A fire occurred in a residential building in Dijon, France on 14 November 2010 and resulted in 7 fatalities and 11 injuries [42]. The fire started in an external garbage container at the base of the building resulting in rapid vertical fire spread on the

**Fig. 5.6** Dijon, France fire 2010 [42]

façade. The façade is believed to be EIFS system with EPS insulation and mineral wool fire barriers, however no detailed fire investigation reports have been found. Significant smoke spread through the building was reported with 130 occupants being evacuated and some occupants jumping from windows. The wind was reported to be blowing the flames against the wall. From the image below it appears that much of the vertical fire spread appears to be centred on a vertical "U" shaped channel profile created by balconies (Fig. 5.6).

## 5.1.5 Berlin, Germany, 2005 [43]

A fire occurred in a seven storey apartment building in Berlin Germany on 21 April 2005. The building was constructed in 1995–1996. The external walls were constructed of concrete poured with a lost formwork consisting of 25 mm chipboard (which remains in place after construction). Eighty millimeter thick EPS foam insulation was fixed directly to the chipboard and was encased in a mesh and render. In 2004 a 500 mm thick fire barrier (mineral fibre) was added to the second and fourth levels.

The fire started at 1.50 am in a second floor apparent. The resulting room fire flashed over with flames extending from the window. This resulted in ignition and fire spread vertically up the EIFS to the top of the building. It is estimated that the time from the initiating room fire to spread to the top of the building via the façade was approximately 20 min.

Some fire spread into the rooms on levels above was reported with significant smoke spread through the entire building. The fire resulted in two fatalities and three people injured.

The façade consisted of 80 mm flame-retarded expanded polystyrene (EPS) with mesh and render and mounted on 25 mm thick chipboard, which was the formwork left in place when the concrete walls had been built (Fig. 5.7).

**Fig. 5.7** Berlin EIFS fire 2005 [43]

**Fig. 5.8** Munich EIFS fire
1996 [1]

## 5.1.6  Apartment Building, Munich, 1996 [1, 44]

A five-level apartment building with a façade made of a composite thermal insulation (about 100 mm thick) comprising polystyrene and foam plastics slabs and a
reinforced covering layer. A rubbish container fire on the exterior ignited the cladding and created extensive damage. Windows were broken and flames spread into
rooms at upper levels (Fig. 5.8).

## 5.2   Fires Involving Metal Composite Cladding

### 5.2.1   Mermoz Tower, Roubaix France, 2012 [45–48]

Mermoz Tower is an 18 storey residential building in Roubaix, France. The building was refurbished in 2003. The refurbishment included installation of metal composite cladding to the middle part of the façade, see Fig. 5.9. This included the exterior walls within balconies, Fig. 5.10.

On the first storey only, "formo-phenolic" decorative boards were installed. On all the other 17 storeys metal composite cladding consisting of a 3 mm

**Fig. 5.9** Metal composite cladding on exterior of Mermoz Tower [45]

**Fig. 5.10** Metal composite cladding on exterior walls of balconies [45]

**Fig. 5.11** Image from building next door to Mermoz Tower refurbished with the same materials. "Formo-phenolic" board shown at *bottom* and metal composite cladding shown at *top* [45]

thick polyethylene core sandwiched between two 0.5 mm thick aluminium sheets (Fig. 5.11).

On 14 May 2010 a domestic fire on a second storey balcony. This resulted in rapid vertical flame spread to the top of the building within a few minutes. Video of the fire shows that the fire spread appeared to be enhanced by the vertical "U" shaped channel profile created by the balconies, with flames moving in-and-out of balconies on each level as the fire spread upwards. Windows on the exposed façade were broken resulting in smoke filling into the building interior. Video also shows molten flaming debris from the façade panels falling to the ground and lower level balconies. This fire resulted in one fatality and six injuries (Fig. 5.12).

## 5.2.2  Al Tayer Tower, Sharjah, 2012 [49, 50]

Al Tayer Tower is a residential building with 408 apartments, 34 residential floors and 6 parking storeys. The exterior of the building was clad with metal composite panels consisting of aluminium with a polyethylene core. On April 28 2012, fire

**Fig. 5.12**  Mermoz Tower during and after façade fire [45]

**Fig. 5.13**  Al Tayer Tower façade fire 2012 [49, 50]

started on a balcony on the first floor. The fire is believed to have started from discarded cigarette landing on the balcony which contained cardboard boxes and plastics. This resulted in vertical fire spread on the metal composite cladding to the top of the building. It also resulted in damage to 45 vehicles parked near the building due to burning falling debris. Newspaper reports also state that this resulted in a significant housing shortage for displaced occupants. No deaths or injuries reported (Fig. 5.13).

### 5.2.3   Saif Belhasa Building, Tecom, Dubai 2012 [51, 52]

The Saif Belhasa building is a 13 storey residential building with 156 apartments and lower level car parking. It is located in the Tecom district in Dubai. The building was clad with metal composite panels consisting of aluminium with a polyethylene core. On 6 October 2012 a fire started on the fourth floor. The fire rapidly spread to reach the top of the building. This resulted in at least two injuries, nine separate flats and their contents were destroyed and at least five cars parked at street level below were damaged by falling burning debris. Fire fighting teams including a truck with crane were dispatched at 9.35 am and the fire was suppressed by 10.57 am. Photos appear to show that vertical spread was centred on vertical channel profiles created by balconies (Figs. 5.14 and 5.15).

### 5.2.4   Tamweel Tower, Dubai, 2012 [53–55]

The Tamweel Tower is a 34 storey mixed use and residential building located in Jumeirah Lakes, Dubai. The building was clad with metal composite panels consisting of aluminium with a polyethylene core. The metal composite cladding was also used as a decorative feature on the roof top. On 18 November 2012 at 1.30 am a fire started at the roof level, possibly near air conditioning equipment. The fire then spread down the exterior of the building. Based on photos and video it appears that the downward fire spread was at least partially due to molten flaming debris from the cladding falling onto lower level balconies and igniting the façade at lower levels. The fire was suppressed by fire brigades at around 8.20 am. No Fatalities were identified in reports reviewed (Fig. 5.16).

**Fig. 5.14**  Saif Belhasa building façade fire 2012 [51, 52]

**Fig. 5.15** Saif Belhasa building façade fire, burning debris falling to base of building [47]

**Fig. 5.16**  Tamweel Tower fire 2012 [55]

### 5.2.5   *Wooshin Golden Suites, Busan South Korea [56–59]*

The Wooshin Golden Suites in the Haeundai district, Busan, South Korea is a, 140 m high mixed use (mostly apartment) building. The building construction was completed in December 2005. It had a steel structure with reinforced concrete structure in part. The building had 38 stories above ground and 4 stories underground and had a total floor area of 68,917 m$^2$. The first floor was for commercial/retail use, the second and third floors were shared facilities including gym, pool and meeting rooms. The 4th floor was a plant and equipment level and floors 5–38 were mostly residential with some office.

The building was constructed with a curtain wall façade with metal composite panels consisting of aluminium with a 3 mm polyethylene core. As the name indicates the panels were gold in colour. The fire was examined in detail in a "fire science and technology" journal article [58]. This presented the following cross section of the façade which indicates glass wool thermal insulation. However some newspaper articles indicated that the thermal insulation may have been EPS (Fig. 5.17).

The fire is reported to have started on the fourth floor due to a spark from an electrical outlet igniting nearby objects. The building was reported to be sprinkler protected but not to have sprinklers in the room of fire origin. The fire spread to the exterior façade and then spread vertically upward on the façade reaching the top of the building within 20 min, destroying the roof top sky lounge, penthouse and some units on the 37th floor. The vertical fire spread was centred around a vertical "U" shaped channel in the external profile of the building (near the central stairways). This appeared to enhance the fire spread through re-radiation and chimney effect. The fire spread may also have been enhanced by the strong wind blowing in from the sea with wind impinging on the side of the building an blowing up thought the "U" shaped external profile. The fire brigade use helicopters to evacuate some occupants from the roof and also to water bomb the exterior of the building from the air. The fire in the room of fire origin was suppressed by 1 pm and the fire for the entire building was suppressed by 6.48 pm. The total interior fire spread floor area was 1134 m². Only four injuries were reported with no fatalities. The financial loss is estimated to be more than 400 million Yen (Fig. 5.18).

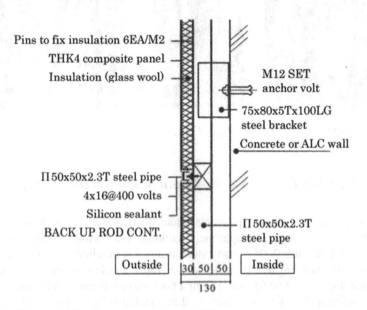

**Fig. 5.17** Wooshin Golden Suites curtain wall detail [59]

**Fig. 5.18**   Wooshin Golden Suites Fire [59]

### 5.2.6   Water Club Tower at the Borgata Casino Hotel, Atlantic City, USA [60]

On September 23, 2007 an external wall fire occurred on the Water Club Tower at the Borgata Casino Hotel, Atlantic City. The building was under construction and nearing completion. The building was 41 stories and was clad with an aluminium composite panel system having a polyethylene core. The fire started as an internal fire on the third floor. The panels were white in colour and were intended to appear like a sail on the side of the new high-rise tower. Fortunately there was a concrete shear wall 6 ft behind these exterior panels that prevented major fire and smoke spread into the interior of the building. There were no direct openings into the interior portion of the void space other than on the 3rd floor and the roof on the 41st floor. The fire spread vertically and rapidly reached the top of the building on one side of the building. The fire brigade reported that within 10–15 min of their arrival, the bulk of the fire had subsided due to rapid consumption of the available fuel. A significant amount of falling structural debris occurred within about a quarter-mile of the building (Fig. 5.19).

**Fig. 5.19** Water Club Tower fire during and after the fire [60]

## 5.3   Fires Involving Weather Resistive Barriers or Rain Screen Cladding

### 5.3.1   Knowsley Heights, UK, 1991 [61]

Knowsley Heights is an 11-floor apartment building located in Liverpool, UK. Prior to the fire the building was recently fitted with a rain screen cladding installed with a 90 mm air gap behind and rubberised paint coating over the external surface of the concrete wall behind. No fire barriers were provided to the air cavity behind. The rain screen and cavity without barriers covered the building from ground floor to the top of the 11th floor. The rain screen material was a Class O (limited combustibility) rated product using BS 476 parts 6 and 7 (BSI, 1981; BSI, 1987).

A fire started in a rubbish compound outside the building rapidly spread vertically through the 90 mm air cavity. The fire destroyed the rubbish compound and severely damaged the ground floor lobby and the outer walls and windows of all the upper floors. No smoke or fire spread to the interior of the building.

BRE has cited this fire incident as a motivator for subsequent building code changes and development of a large scale façade fire test (Fig. 5.20).

**Fig. 5.20**  Knowsley Heights fire 1991 [62]

## 5.4   Fires Involving Insulated Sandwich Panels

### 5.4.1   Tip Top Bakery Fire, NSW, Australia 2002 [63]

The Tip Top Bakery was located in Fairfield, NSW, Australia. It was a single level large factory with a floor area of 10,000 m$^2$. The walls and in some areas the roof were constructed of polystyrene insulated sandwich panels. The building was not sprinkler protected but was provided with a thermal fire detection system connected to fire brigade monitoring. The fire occurred on 2 June 2002. The cause of the fire was determined to be failure of a gas fired heating system resulting in ignition of polenta flour. The fire brigade initially commenced internal offensive fire fighting, however due to poor water supply, rapid fire spread and identification of the EPS sandwich panels the fire brigade switched to Defensive fire fighting. The fire resulted in destruction of most of the building and a total loss in excess of $100 Million. The fire brigade incident report highlights the risk of (and observed resulting) structural collapse of the EPS sandwich panels as a major factor in switch from offensive to defensive fire fighting (Fig. 5.21).

### 5.4.2   UK Sandwich Panel Fire Incidents

Harwood and Hume [64] report on an investigation of 21 fire incidents involving sandwich panels by the Fire Research Station (FRS). Two incidents involved purely cold storage buildings, 12 of the incidents involved food processing plants and a further 5 incidents were in factory buildings.

**Fig. 5.21**  Tip Top Bakery Fire, 2002 [63]

The FRS study identifies that small fires are not uncommon in food processing plants and are routinely extinguished by staff. However, if staff are not present when a fire starts, or the fire is hidden or the cause is not routine, it is then more likely to develop and spread into the sandwich panels, with the possible loss of the factory. All the fires that involved sandwich panels produced large quantities of black smoke. In many cases fire fighters needed to use breathing apparatus while working around the outside perimeter of the building.

In eight incidents the fire brigade was unable to carry out fire fighting within the building and in another three they were forced to retreat from the building. Two fire fighters died in the Sun Valley Poultry fire in Hereford in 1993, trapped by the collapse of panels. No fatalities were reported for the rest of the incidents reviewed. However, other brigades also report panels collapsing as they retreated out of the building or fought the fire from the entrances. In all cases investigated, the occupants had left the building safely before the fire had developed sufficiently to put them at risk.

## 5.5   Fires Involving Other Types of Exterior Systems

### 5.5.1   Apartment Building, Irvine, Scotland, 1999 [1, 65]

Windows at the corners of a 13-storey apartment tower in Irvine, Scotland, had been letting in cold and/or moisture. In order to eliminate these problems and also to improve visual appearance, new window frames of unplastisized polyvinyl chloride (uPVC) were fixed. The exterior wall around the window was covered with glass reinforced polyester plastic sheet. This gave a picture frame effect around the window. The glass reinforced polyester sheet was also extended below the window.

On 11 June 1999 a fire started in a room on the fifth floor. The fire burnt out through and window and with approximately 10 min had spread vertically up seven floors to the top of the building. The fire spread was limited to the strip of external

**Fig. 5.22**   Irvine Scotland Fire 1999 [1]

combustible materials about the windows. It is not clear if the fire was spread by means of the surface of the plastic sheet or whether the fire spread within a cavity that may have existed between the cladding and the original external wall. There was one fatality; a wheelchair bound man in the apartment of fire origin (Fig. 5.22).

## 5.5.2   Eldorado Hotel, Reno, Nevada 1997 [66, 67]

A large fire occurred on the façade of the Eldorado Hotel, Reno on September 30 1997. The Façade is reported as being "plastic" in newspaper reports. The material was believed to be hard coat polyurethane over EPS and was believed to extend approximately 120 ft long and 60 ft high but no detailed description of the façade in included in reports. The fire was reported to have started at 6.40 pm due to an electrical fault on the external wall. Flames reached up to 50 m above the second floor roof. The fire was extinguished by the fire brigade within 45 min. The fire did not spread to the building's interior. The building was sprinkler protected but no internal sprinklers are reported to have activated (Fig. 5.23).

## 5.5.3   Palace Station Hotel and Casino, Las Vegas, USA 1998 [67, 68]

Palace Station is a 20 storey hotel. A fire started on the external façade at the top of the building at approximately 6.30 am June 20 1998. The fire was confined to the outside of the building at the 20th floor and roof. An external decorative façade was

**Fig. 5.23** Eldorado Hotel
Fire 1997 [66, 67]

**Fig. 5.24** Palace Station Hotel fire 1998 [68]

the only object that burned. A 12-21-99 letter from Don Belles to CCBD specifies that polyurethane foam and urethane coated EPS was used [67]. The fire did not spread to the interior of the building. The fire is believed to have been caused by a lightning circuit on the outside of the façade (Fig. 5.24).

### 5.5.4   Grozny-City Towers, Chechnya, Russia, 2013 [69–71]

On April 3, 2013 a fire occurred on the façade of the Grozny-City tower (Fig. 5.25).

The building was a 145 m high, 40-storey high rise building that was unoccupied. It had just completed construction and may have had final construction works underway. The fire is thought to have started due to a short circuit in an air conditioner on the upper floors. The fire systems for the building had not been commissioned and there appeared to be no water supply to sprinklers or hydrants. The fire spread to engulf 18,000 m$^2$ of the façade from ground level to the roof. No details of the façade material are reported other than being "plastic insulating plates." Based on photos and videos the material may have been a metal composite panel but this has not been verified. The fire took 8 h for fire brigades to extinguish.

### 5.5.5   Fire Incidents Reported in China

With China's social and economic development and urbanization, the number of (ultra) high-rise buildings is increasing. Approximately 200,000 high-rise buildings, in which 3000 are ultra-high-rise buildings, are found in Mainland China. It can be deduced from reported fire incidents that the rapid development of high-rise buildings may have resulted in poor regulation of combustible exterior materials which poses serious fire safety problems.

**Fig. 5.25** Grozny-City
Tower fire, 2013 [70]

Since 2006, several large fire incidents have occurred in high-rise buildings (see figures below). These have included the Central Television headquarters (CCTV Tower) Fire in 2009 (February 9), resulting in one fire fighter's death, seven people injured, and direct economic loss of 1.6 billion RMB. A 44 storey tower nearing completion of construction. The facade at the top of the building was ignited by illegal fireworks. The fire spread to involve the majority of the facade over the entire height of building. The façade is believed to have included a polystyrene insulation [72].

An exterior façade fire occurred in a 28-storey residential building in Shanghai Jing'an District on November 15 2010 killed 58 people, and injured over 70 people [62]. This fire was believed to be caused by welding resulting in fire spread on polyurethane insulation to external walls. Other large fires include the Harbin "Jingwei 360 degrees" building fire in 2008 (not shown) and the Shenyang Royal Wanxin building fire in 2011, a catastrophic fire. All these high-rise building fires have caused a large amount of economic and property loss, resulted in significant social impacts, and highlighted the research needs of fire safety in high-rise buildings. Unfortunately, detailed information on these incidents or any regulatory changes in China has not been available (Figs. 5.26 and 5.27).

### 5.5.6  Fire Incidents in Japan

It was not possible to find several external façade incidents in Japan. It seems that such incidents listed as very large scale building fires may have occurred prior to 1974 after which stringent fire regulations required fire resistance construction of the buildings [73].

**Fig. 5.26**  Shanghai Fire (*left*) [62] and CCTV Tower, Beijing fire (*right*) [72]

研究背景

建筑外立面火灾的严重性

2012.12.22 Yan' an

2011.2.3 Shenyang

**Fig. 5.27** Façade fire incidents in China, kindly provided by SKLFS (State Key Laboratory for Fire Science) at the USTC (University of Science and Technology China) University of China. No detailed information is available

The only recorded external facade fire incident occurred in 1996 and caused fast fire spread owing to PMMA fences used in the balconies [74]. Vertical Fire Spread along Balcony of High-rise Apartment in Hiroshima Motomachi. On 28 October 1996, there was a massive façade fire at high-rise apartment in Hiroshima Motomachi, Japan. Fire started on the 9th floor and vertically propagated along acrylic blindfold (PMMA) boards of balcony, to the 20th floor (top floor). It was very rapid. The fire reached the top floor within 10 min, which demonstrated the potential danger of combustible components located at the exterior façade of buildings.

## 5.6   Summary of Observations from Case Studies

- Although exterior wall fires are low frequency events, the resulting consequences in terms of extent of fire spread and property loss can be potentially very high.
- For most of the incidents reviewed the impact on life safety in terms of deaths has been relatively low with the main impacts being due to smoke exposure rather than direct flame or heat exposure. However a large number of occupants are usually displaced for significant periods after the fire incidents.
- Fire incidents appear to predominantly have occurred in countries with poor (or no) regulatory controls on combustible exterior walls at the time or where construction has not been accordance with regulatory controls.

- Internal fires which spread to the exterior wall are the most common fire start scenario for the incidents reviewed.
- Falling burning debris can be a significant hazard relating to these fires and causes downward fire spread.
- Re entrant corners and channels that form "chimneys" has lead to more extensive flame spread than flat walls. The affect of balconies forming partial vertical "channels" should be further investigated.
- Combustible exterior wall systems may present an increased fire hazard during installation and construction prior to complete finishing and protection of the systems. The 2009 CCTV Tower Fire and 2010 Shanghai fire in China are examples of large fires occurring during construction.

# Chapter 6
# Regulation

A range of regulations and building codes around the world have been reviewed and a detailed summary of the prescriptive requirements relating to combustible exterior wall assemblies is provided in Appendix B.

The following key aspects of regulation have been identified to have significant impact on performance of exterior wall assemblies and fire risk and therefore the review has focussed primarily on these aspects:

1. Reaction to fire requirements for exterior wall assemblies and materials
2. Fire stopping/barrier requirements both in and behind exterior walls
3. Separation of buildings, in terms of minimum separation of unprotected openings from a relevant boundary.
4. Separation of openings between stories
5. Requirements for sprinkler protection—which influences the risk of an initiating compartment fire and fire spread into compartments

The regulations reviewed range from purely prescriptive code requirements to performance based codes which enable either a prescriptive solution or an alternative solution which must be justified based on fire engineering analysis. However this review has focused primarily on the prescriptive requirements.

Fire engineering modelling and analysis in the area of fire spread on combustible material assemblies is not as well established or reliable as other areas such as smoke behaviour or evacuation movement. Therefore it would be best practice to base any fire engineering analysis of a performance based design (Alternative solution) relating to combustible exterior wall assemblies on full-scale testing and risk assessment.

© Fire Protection Research Foundation 2015
N. White, M. Delichatsios, *Fire Hazards of Exterior Wall Assemblies Containing Combustible Components*, SpringerBriefs in Fire, DOI 10.1007/978-1-4939-2898-9_6

## 6.1   Reaction to Fire Requirements

Reaction to fire refers to the requirements for ignition, combustibility and fire spread on assemblies or individual materials. The various regulations around the world generally fit into one of four categories

1. No Reaction to fire requirements,
2. Requirement for non-combustible materials only for over 4 storeys.
3. Requirements for small scale reaction to fire tests only
4. Requirement for Full-scale Façade test, in some cases full scale test not required for specific types of materials meeting small-scale test requirements.

### 6.1.1   Australia and New Zealand

The Australian National Construction Code [75, 76] does not state any requirements other than non-combustibility. Residential or public assembly buildings of two stories or more and all other classes of buildings of three stories or more are not permitted to have combustible external walls. In practice, combustible external wall assemblies are often used for buildings greater than three stories via performance based fire engineered alternative solutions. However the lack of any full-scale façade testing in this country sometimes results in fire engineers and certifiers accepting materials based on very limited small scale tests (and in some cases qualitative risk assessment with no tests at all).

The New Zealand building code [77] regulates external wall assemblies based on peak HRR and total heat released in cone calorimeter testing. Alternatively compliance with NFPA 285 or "other full-scale façade tests" may be used. These requirements generally apply to buildings greater than 7 m high or less than 1 m from a relevant boundary. If buildings are less than 25 m high and sprinkler protected then there are no requirements for combustible exterior wall materials.

### 6.1.2   UK

The UK Building Regulations and Approved Document B [78] requires either compliance with BRE Report BR135 using full scale façade tests BS8414 Part 1 [79] or Part 2 [80], or requires materials to be non-combustible or limited combustibility materials based on either BS 476 Part 6 [81] and Part 11 [82] tests or Eurocode classification [83] (Class B-s3,d2 or better). These requirements apply to buildings 18 m or more high or less than 1 m from a relevant boundary.

However often in the UK insurers require compliance with Loss Prevention Standard LPS 1181 Part 4 [84] which requires compliance with BRE Report BR135 and full-scale testing to BS 8414 (which eliminates acceptance on small scale tests alone). This LPS standard also requires cone calorimeter testing on combustible components for quality control.

### 6.1.3 Nordic Countries and Europe

The reaction to fire requirements for exterior wall materials in Nordic countries are generally based on Euroclassifications. Acceptable solutions vary from non-combustible materials (A2-s1,d0) to only fulfilling variations of Euroclass B. In Sweden, full-scale testing to SP Fire 105 is also accepted as an alternative. Some countries allow some parts of the façade to be of a lower class, i.e. D-s2,d0 [85].

European countries such as France and Germany generally apply Euroclassifications or alternatively full-scale façade tests.

### 6.1.4 USA

In the USA the International Building Code (IBC) [86] is the model building code adopted by most states. NFPA 5000 [87] is an alternative building code to the IBC but it is not adopted by most states. There are some detailed differences in requirements for exterior walls between these two codes however they are similar in terms of the types of testing that are required.

The IBC requires compliance with the full-scale test NFPA 285 [88] for buildings greater than 12.192 m in height. However there are number of specific exceptions to permit different types of materials without full-scale tests based on small scale tests, mainly the ASTM E84 [89] or UL 723 [90] flame spread test and the ASTM D 1929 [91] ignition temperature test. NFPA 5000 generally requires compliance with the full-scale test NFPA 285, regardless of height. However there are specific exceptions to permit different types of materials without full-scale tests based on the same small scale tests as the IBC (for example metal composite panels installed to a maximum height of 15 m).

The exceptions permitting small scale testing rather than full-scale testing in the IBC and NFPA 5000 are complex to read and understand which could possibly lead to miss-interpretation and poor compliance.

FM Global insurer requirements refer to FM 4880 tests [92], including room corner tests, parallel panel tests, and 25 and 50 ft corner tests. In practice these are mainly applied to Insulated Sandwich Panels for industrial and storage type buildings. These requirements are applied to countries beyond the USA where FM Global is an Insurer.

### 6.1.5   Canada

The National Fire Code of Canada [93] requires full-scale façade testing to CAN/ULC S134 [94].

### 6.1.6   United Arab Emirates

Prior to 2012 the UAE Fire & Safety Code [95] did not set any requirements for reaction to fire of exterior wall materials. In response to a spate of large fire incidents predominantly involving metal composite materials, Annexure A.1.21 of the UAE fire & life safety code [96] was released which provides specific requirements for reaction to fire of exterior wall cladding and passive fire stopping.

For buildings 15 m or greater in height or less than 3 m from a boundary compliance with BR135 and full-scale façade testing to BS 8414 Parts 1 or 2 in addition to a range of possible small scale US, BS or Euroclass tests or room corner tests are required. For buildings less than 15 m high but less than 3 m from a boundary compliance with only the small scale or room corner tests is requires.

### 6.1.7   Singapore and Malaysia

Singapore [97] and Malaysian [98, 99] building regulations have been developed based on early UK regulations but have not been updated to include full-scale façade test requirements. In both countries exterior wall cladding must be either non combustible or Class 0 materials (Flame spread index $\leq 12$ and sub index not exceeding 6 when tested to BS 476 Part 6).

In Singapore this requirement is applied to buildings greater than 15 m high or less than 1 m from a boundary. In Malaysia this requirement is applied to buildings greater than 18 m high or less than 1.2 m from a boundary.

### 6.1.8   China and Japan

China applies a full-scale façade test method which is similar in its dimensions and measurements to BS 8414-1.

We note that according to current building regulations in Japan for fire protection for exterior walls of a building, only fire resistance performance of the wall system is considered even when the facades are combustible. No consideration is given on the evaluation of ignition and fire propagation on combustible facades. For example, in case of exterior walls for reinforced concrete structures it is considered acceptable to have attached foamed plastics as insulating materials. Standards for façade flames on non-combustible or combustible exterior wall systems do not exist.

To address this situation, a new facade test is currently proposed in Japan, see Sect. 7.3.2.

## 6.2   Fire Stopping

Most countries reviewed require fire stopping to gaps at the rear of the external wall at the junction of floors or compartment boundaries (i.e. curtain wall fire stopping). This fire stopping is generally required to have a fire resistance rating equivalent to the surrounding construction.

In some cases such as the UK, the USA ICC, New Zealand and the UAE fire stopping to limit the size of cavities in exterior walls is required, however, in other countries such as Australia there no stated requirement for this.

Fire stopping imbedded in EIFS around openings and at set intervals is not generally explicitly stated for most countries however may be required to pass Full-scale façade tests in countries where this is required.

EIFS standards and guidelines including ASTM E2568 [100], EIMA 99A [101] and ETAG 004 [102] do not explicitly require fire stopping imbedded in EIFS but do require re-forced rendering to cover off openings such as windows etc.

## 6.3   Separation of Buildings

For the majority of countries reviewed the minimum separation of a building from a boundary permitted with unprotected openings was 1 m (NZ, UK, Malaysia, Singapore, USA sprinklered), 1.5 m (USA non sprinklered) or 3 m (Australia, UAE). As separation distance increases beyond these minimum distances the percentage area of unprotected openings gradually increases.

In Australia the minimum separation distance permitted with unprotected openings is 3 m. At separation distanced greater than 3 m the area of unprotected openings is unrestricted.

## 6.4   Separation of Openings Vertically Between Stories of Fire Compartments

Almost all of the countries reviewed required separation of unprotected openings vertically between stories or fire compartments for buildings taller than 3–4 stories by either

- A 915 mm spandrel with 1 h fire resistance
- A 760 mm horizontally projecting barrier with 1 h fire resistance.

In some countries the above is not required where the building is sprinkler protected. In some countries this concession for sprinkler protected buildings is not provided.

## 6.5  Sprinkler Protection

The requirements for minimum building height at which sprinkler protection is to be provided varies considerably between countries ranging from 16.8 m (USA ICC) 22.9 m (USA NFPA 5000), 24 m (Singapore) 25 m (Australia, New Zealand), 30 m (UK).

In some countries the minimum height requiring sprinkler protection varies with the type of building use (UAE and Malaysia).

In the USA all residential apartments generally require sprinkler protection.

Sprinkler protection is also required when fire compartment size limited specified for each country are exceeded.

## 6.6  Discussion

Of the five aspects of regulation reviewed, the reaction to fire regulation requirements are expected to have the most significant impact on actual fire performance and level of fire risk presented by exterior wall assemblies.

In Australia where there are non-combustibility requirements, combustible materials are often applied as Alternative solutions with varying levels of relevant test based evidence of performance.

In New Zealand the use of the small scale Cone calorimeter to regulate exterior wall assemblies may not be correctly characterising the full-scale complex performance of exterior wall assemblies.

Countries such as the USA, UK, and some European countries specify full-scale façade testing but then permit exemptions for specific types of material based on small-scale fire testing. The correlation between these small scale tests and full-scale performance is not fully understood for the range of exterior wall materials and assemblies they are applied to.

The correlation between small scale tests and full-scale performance needs further research to enable cost effective regulation.

# Chapter 7
# Test Methods

## 7.1 Full-Scale Façade Fire Test Methods

The following describes some of the main full-scale façade fire test methods worldwide. Please refer to Appendix C for a table which summarises the full-scale and intermediate scale façade fire test methods reviewed.

### 7.1.1 ISO 13785:2002 Part 2 [12]

The two parts of ISO 13785 provide two test methods:

- Part 1 is an intermediate scale test intended as a less expensive screening test for product developers to assess and eliminate materials or sub-components which fail prior to undertaking a full-scale test (described in Sect. 7.2.1).
- Part 2 is a full-scale faced test (described in this section)

These tests are applicable only to façades and claddings that are non-load bearing. No attempt is made to determine the structural strength of the façade or cladding under fire conditions.

For ISO 13785-2 the test façade is installed as a re-entrant corner "L" arrangement or wing wall. The fire source is flames emerging from a compartment fire via a window. The height of the tested façade is at least 4 m above the window lintel. The main façade is at least 3 m wide and the wing façade is at least 1.2 m wide. The window is on the main wall with one edge at the wing wall and is 2 m wide × 1.2 m high. The façade is installed around the window down to the bottom of the window. The façade is installed representative of the end use including all insulation, cavity air gaps, fixings and window details.

© Fire Protection Research Foundation 2015
N. White, M. Delichatsios, *Fire Hazards of Exterior Wall Assemblies Containing Combustible Components*, SpringerBriefs in Fire,
DOI 10.1007/978-1-4939-2898-9_7

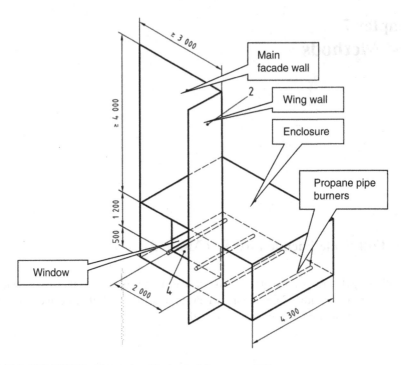

**Fig. 7.1** ISO 13785 Part 2 test rig with standard fire source [12]

The fire source is located within a fire enclosure and may be any source which is calibrated to achieve an average total heat flux of $55 \pm 5$ kW/m² at a height of 0.6 m above the window and an average total heat flux of $35 \pm 5$ kW/m² at a height of 1.6 m above the window. The fire source has a 4–6 min growth phase and a similar decay phase. The total test duration is 23–27 min. The standard fire source is series of large perforated pipe propane burners installed in an enclosure approximately 4 m wide × 4 m deep × 2 m high with a total output of 5.5 MW. Alternative fire sources are permitted and the fire enclosure may any volume in the range 20–30 m³.

During the test total heat flux is measured across the façade surface at 0.6, 1.6 and 3.6 m above the window. Thermocouples are located on the outside surface of the façade immediately above the window and also at 4 m above the window. Thermocouples are also inserted into intermediate layers of material and cavity air gaps at height of 4 m above the window. Fire spread is observed (Fig. 7.1).

Potential problems with this test method may include the significant space and gas supply required for the large standard enclosure and fire source. The use of permissible alternative enclosure sizes and fire sources may alleviate this. However this results in a lower intensity fire exposure to the façade when compared with some other large scale tests. Also the allowable variance of the fire source including the growth and decay times may result in some variance to test exposures.

## 7.1.2   BS 8414 Part 1 and Part 2 [79, 80]

BS 8414 part 1 and part 2 were developed by BRE. BS 8414-1 is a full-scale fire test for non-load bearing external cladding systems applied to the face of a solid external building wall. The test simulates the scenario of flames emerging from a compartment fire via a window at the base of the wall. The test façade is installed as a re-entrant corner "L" arrangement. The test rig has a masonry block wall construction as the substrate for mounting test specimens to. The test wall extends at least 6 m above the window soffit. The main wall is at least 2.6 m wide and the wing wall is at least 1.5 m wide. The window opening is at the base of the main wall and is 2 m wide × 2 m high. The façade is installed around the window down to the bottom of the window. The façade is installed representative of the end use including all insulation, cavity air gaps, fixings and window details. The tested façade must be at least 2.4 m wide on the main wall and 1.2 m wide on the wing wall (Fig. 7.2).

The fire enclosure is 2 m wide × 1 m deep × 2.23 m high with a lintel at the front opening reducing the soffit height of the opening to 2 m. The standard fire source is a timber crib constructed of softwood sticks having a cross sectional area of 50 mm × 50 mm. The constructed timber crib is nominally 1.5 m wide × 1 m

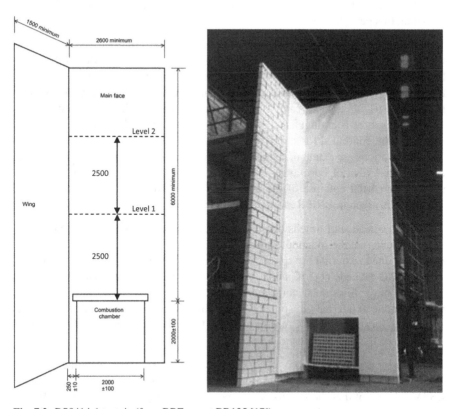

**Fig. 7.2** BS8414-1 test rig (from BRE report BR135 [17])

deep × 1 m high. The crib sits on a platform 400 mm above the base of the test frame and the front of the crib sits 100 mm in front of the outside surface of the masonry support wall. Therefore the front of the crib is directly 600 mm under the soffit of the tested façade. The crib has a nominal heat output of 4500 MJ over 30 min and a peak HRR of $3 \pm 0.5$ MW. The standard fire source achieves the following calibrated exposure.

- The mean temperature across the top of the combustion chamber opening measured at three thermocouple locations exceeds 600 °C above ambient over a continuous 20 min period. The variation between mean temperature and any individual thermocouple temperature shall not exceed ±20 °C
- The mean temperature at level 1 height on the main wall face exceeds 500 °C above ambient over a continuous 20 min period.

Alternative fuel sources such as gas burners can be used but must achieve the above temperature exposure and the following additional heat flux requirement for fuels other than cribs:

- Mean heat flux measured at 1 m above the window soffit on the main wall shall remain within the range of 45–95 kW/m$^2$ over a continuous 20 min period and typically achieves a steady state peak mean heat flux of approximately 75 kW/m$^2$ within this period.

During the test temperatures are measured at the external surface at the test façade on the main and wing walls at level 1 (2.5 m above the window soffit) and level 2 (5 m above the window soffit). Internal thermocouples are only located at level 2 on the main and wing wall and are positioned at the centre of each combustible layer >10 mm thick or cavity. No heat flux is measured during the test.

The fire source is extinguished 30 min after ignition and observations and measurements are continued for a total test period of 60 min or until all flaming ceases. Key observations are extent of flame spread on all surfaces, intermediate layers and cavities, the extent of burn away or detachment for the cladding system and any collapse or partial collapse of the cladding system. The performance criteria for BS8414-1 is given in BRE Report BR135 [17] and is:

- The fire spread start time is defined as the time when the temperature measured by any external thermocouple at level 1 exceeds 200 °C above ambient
- Failure due to external fire spread is determined when any external thermocouple at level 2 exceeds 600 °C above ambient for a period of at least 30 s, within 15 min of the fire spread start time
- Failure due to internal fire spread is determined when any internal thermocouple at level 2 exceeds 600 °C above ambient for a period of at least 30 s, within 15 min of the fire spread start time

BS8414-2 is a full-scale fire test for non-load bearing external cladding systems fixed to and supported by a structural steel frame. This test is essentially the same as BS8414-1 except that the test façade is mounted directly to a steel support frame without the masonry substrate. This tests curtain wall type construction where a

**Fig. 7.3** BS8414-2 test rig (from BRE report BR135 [17])

solid concrete or masonry wall is not present. The dimensions of the test rig, the fire source and the test procedure are the same as for BS8414-1. The performance criteria for BS8414-2 is given in BRE Report BR135 [17] and is the same as for BS8414-1 except for the following additional criteria for internal fire spread.

- Failure due to internal fire spread is also determined when burn through of the façade system with continuous flaming with a duration of at least 60 s is observed on the non-exposed side of the facade at a height of 0.5 m or greater above the window soffit within 15 min of the fire spread start time (Fig. 7.3).

There are no failure criteria set for mechanical performance by the BS8414 standards or the BRE report BR135. However observation of mechanical behaviour including system collapse, spalling, flaming debris etc. should be recorded.

### 7.1.3   DIN 4102-20 (Draft)

Please note that Authors have not had access to DIN 4102-20 (Draft). The following description has been determined from descriptions provided in other reports [103].

This test simulates the scenario of flames emerging from a compartment fire via a window at the base of the wall. The test façade is installed as a re-entrant corner "L" arrangement. The test rig has a light weight concrete wall construction as the substrate for mounting test specimens to. The test wall extends at least 5.5 m high. The main wall is at least 2 m wide (using the burner) or 1.8 m wide (using the crib) and the wing wall is at least 1.4 m wide (using the burner) or 1.2 m wide using the crib. The fire enclosure and opening is nominally 1 m wide × 1 m high and is located at the base of the main wall at the intersection of the wing wall. The façade is typically installed around the opening down to floor level. The façade is installed representative of the end use including all insulation, cavity air gaps, fixings and window details.

The fire source is a 320 kW constant HRR linear gas burner located approximately 200 mm below the soffit of the opening. A 25 kg timber crib, 0.5 m × 0.5 m × 0.48 m, using 40 mm × 40 mm softwood sticks was previously used as the standard fire source. The fire source achieves a maximum temperature of approximately 780–800 °C measured 1 m above the opening soffit on a non combustible wall. Flames from the fire source are understood to extend a maximum height of 2.5 m above the opening soffit on non-combustible wall.

The gas burner is turned off after 20 min for combustible facades and 30 min for non-combustible facades. Measurements and observations continue until all burning and smoke production ceases, or until 60 min (Fig. 7.4).

The test performance criteria are:

- No burned damaged (excluding melting or sintering) above a height of 3.5 m or more above the opening soffit.
- Temperatures on the wall surface or within the wall layers/cavities must not exceed 500 °C at a height of 3.5 m or more above the opening soffit.
- No observed continuous flaming for more than 30 s at a height of 3.5 m or more above the opening soffit.
- No flames to the top of the specimen at any time.
- Falling of burning droplets and burning and non-burning debris and lateral flame spread must cease with 90 s after burners are turned off.

### 7.1.4   NFPA 285 [88]

This method tests façade claddings or complete external wall systems. The test wall is installed as a single wall surface. No re-entrant corner is installed. The test rig is a two storey steel framed structure with an open fronted test room on each storey constructed of concrete slabs and walls. Each test room has internal dimensions of

**Fig. 7.4** DIN 4102-20
(Draft) test rig (from BRE
Global [103])

approximately 3 m wide × 3 m deep × 2 m high. The bottom test room serves as the
fire enclosure and the top test room simulates an enclosure on the level above with
no window.

The installed test wall is at least 5.3 m high × 4.1 m wide. The wall tested is a
complete system including any external cladding, insulation, external substrate
framing and internal wall membrane. The test wall construction and fastening to the
test rig must be representative of the end use. The test wall is typically installed on
a movable steel frame with is then attached to the front of the test rig concrete slabs.
The test wall includes a single opening 1.98 m wide × 0.76 m high. The opening
soffit is located 1.52 m above the fire enclosure floor.

The fire source consists of two separate pipe type gas burners. One burner is
placed in the centre of the fire enclosure and the other burner in a 1.52 m long linear
burner located near the soffit of the opening. The room burner output is gradually
increased from approximately 690 to 900 kW over the 30 min test duration. The
window burner is ignited 5 min after the room burner and is gradually increased
from 160 to 400 kW over the remaining 25 min test period. The burners are calibrated to achieve average heat fluxes at the surface of a non-combustible test wall of
approximately 40 kW/m² at 0.6 and 0.9 m above the opening and 34 kW/m² at 1.2 m
above the opening during the peak fire source period of 25–30 min.

During the test temperatures are measured at the front of the test wall and also in
air cavity and insulation spaces within the wall at 305 mm intervals vertically from

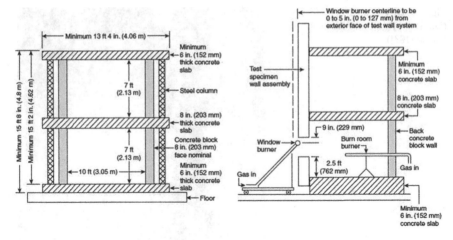

**Fig. 7.5** NFPA 285 test apparatus front view without test wall (*left*) and side view (*right*) (from NFPA 285-2012) [88]

the opening soffit. Temperatures within the fire enclosure, at the rear of the test wall in the second story test room are also measured. No Heat flux measurement is made during the test (Fig. 7.5).

The NFPA 285 standard provides a very detailed set of performance criteria which are briefly summarised as follows.

- Temperatures at exterior of wall must not exceed 538 °C at a height of 3.05 m above the opening soffit.
- Exterior flames must not extend vertically more than 3.05 m above the opening soffit.
- Exterior flames must not extend horizontally more than 1.52 m from the opening centreline.
- Fire spread horizontally and vertically within the wall must not result in designated internal wall cavity and insulation temperatures exceeding stated temperature limits. The position of the designated thermocouples and temperature limits depends on the type and thickness of insulation materials and whether or not an air gap cavity exists.
- Temperatures at the rear of the test wall in the second storey test room must not exceed 278 °C above ambient.
- Flames shall not occur in the second story test room
- Flames must not occur horizontally beyond the intersection of the test wall and the side walls of the test rig.

The NFPA 285 test method is related to a larger façade test developed in 1980 which used a 26 ft (8 m) two story outdoors building. A 1285 lb timber crib was used as the fire source in the lower floor which resulted in flames exiting the window and exposing the exterior face of the wall assembly at approximately 5 min. This test method was published in the 1988 UBC as test standard 17-6 and in the 1994

**Fig. 7.6**   Front view of typical NFPA 285 test (from Hansbro [104])

UBC as UBC test standard 26-4. In the early 1990s a reduced scale, indoors version of the UBC 26-4 test was developed which replaced the wood crib with two gas burners to produce the same exposure. Testing was done to confirm that similar results were achieved for the same materials on the original large and new reduced scale tests. The reduced scale test became UBC 26-9 which eventually replaced UBC 26-4. NFPA 285 is technically equivalent to UBC 26-9 (Fig. 7.6).

### 7.1.5   SP Fire 105 [105]

This method tests façade claddings or complete external wall systems. The wall is installed as a single wall surface 6 m high×4 m wide. No re-entrant corner is installed. The fire source is flames emerging from a compartment fire via a window located at the base of the wall system. The test wall includes two fictitious windows simulating the detail for two above stories. The windows are 1.5 m wide×1.2 m high. A non-combustible eave detail protrudes 500 mm horizontally from the front of the specimen at the top of the test wall.

   The fire enclosure is 3 m wide×1.6 m deep×1.3 m high and has an air intake in the floor. The fire source is a heptane fuel tray 2 m×0.5 m in surface area and 100 mm deep. A calibration test is required to demonstrate that when the heat flux is measured at the centre of the second storey fictitious window, the fire source

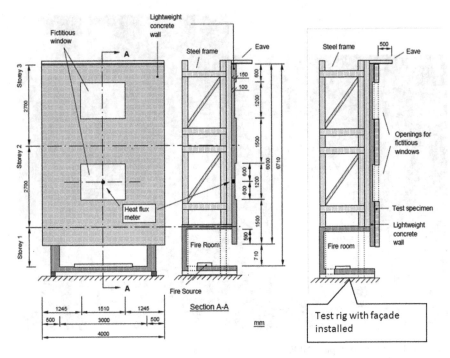

**Fig. 7.7** SP105 test rig [105]

achieves a heat flux of 15 kW/m$^2$ during at least 7 min of the test and 35 kW/m$^2$ during at least 1.5 min of the test and that the heat flux never exceeds 75 kW/m$^2$. Other fire sources are permitted provided the calibration heat flux requirements are met. During the test a minimum of one heat flux meter will be located at the centre of the second storey window and two thermocouples are located to measure gas temperatures at the top of the wall on the underside of the eave. Additional measurements can be made. The test duration is approximately 15 min (Fig. 7.7).

The SP105 standard provides the following performance criteria:

For buildings up to eight stories high which can be reached for external fire fighting, excluding hospitals

(a) No fire spread (flame and damage) higher than the lower part of the second storey window.
(b) No large pieces may fall from the façade.
(c) Temperatures measured at the eave must not exceed 500 °C for more than 2 min or 450 °C for more than 10 min.

For all other buildings including hospitals (a), (b) and (c) above apply. Additionally the heat flux measured at the centre of the window directly above the fire must not exceed 80 kW/m$^2$.

## 7.1.6   CAN/ULC S134 [94]

The Canadian CAN/ULC S134 full scale test method was developed by NRC. The test simulates an enclosure fire exposure via an open window. A single wall surface with no re-entrant corner is tested.

The test method enables complete curtain wall systems to be tested installed to the test support rig with no concrete or masonry substrate if required.

The fire enclosure is 5.95 m wide × 4.4 m deep × 2.75 m high. The window opening is approximately 1.37 m high × 2.6 m wide with the soffit located at the top of the fire enclosure at 2.75 m above floor level.

The total height of the test rig is 10 m. The façade/test wall is installed around the window to a width of 6 m and a height of 7.25 m above the window soffit. The façade is installed representative of the end use including all insulation, cavity air gaps, fixings and window details.

The fire source in the enclosure may be either wood cribs of kiln dried pine with total mass of 675 kg, or four 3.8 m long linear propane burners designed to give the same fire exposure. The burner output is approximately 120 g/s propane (5.5 MW). The total fire exposure time is 25 min with 5 min growth phase, 15 min steady state phase and 5 min decay phase. The fire source is calibrated to achieve a mean heat flux of $45 \pm 5$ kW/m$^2$ measured 0.5 m above the opening soffit, and $27 \pm 3$ kW/m$^2$ measured 1.5 m above the opening soffit when averaged over the 15 min steady state period.

During the test temperatures are measured within the fire enclosure and at the opening 0.15 m below the soffit. Wall temperatures are measured at vertical intervals of 1 m starting at 1.5 m above the opening soffit. At each height temperatures of the front surface and rear surface as well as intermediate material layer and cavity temperatures are measured. Gas temperatures 0.6 m in front of the top of the test wall are also measured. Heat flux at the front surface of the wall 3.5 m above the opening soffit is measured. Radiant heat emitted from the test wall is also measured at a distance of 3 m in front of the wall at heights ranging from 2.1 to 6.0 m above floor level using heat flux meters mounted on a mast (Fig. 7.8).

The performance criterion for this test is specified by the National Building code of Canada which requires:

- Flame spread distance must be less than 5 m above the opening soffit.
- Heat flux measured 3.5 m above the opening soffit must be less than 35 kW/m$^2$

## 7.1.7   ANSI FM 4880 25 and 50 ft Corner Tests [92]

ANSI FM4880 details the FM Approvals process for evaluating insulated wall or wall and roof/ceiling assemblies, plastic interior finish materials, plastic exterior building panels, wall ceiling and coating systems and interior or exterior finish systems. Part of this evaluation process details

**Fig. 7.8** Typical CAN/ULC
S134 test 15 min after
ignition (from NRC test
report [61])

- A 25 ft high corner test to be applied for acceptance of assemblies for an end use
  maximum height of 30 ft (9.1 m)
- A 50 ft high corner test to be applied for acceptance of assemblies for an end use
  maximum height of 50 ft (15.2 m) or unlimited height

Although ANSI FM 4880 states that it is applicable for exterior finish systems, the use of the above two tests is mostly applied to assessing insulated sandwich panels, however FM-Global has done some work assessing other façade materials including EIFS. These tests and are not specifically external façade tests and are not referred to by building codes for regulation of external facades. However these test methods are summarised here as they do provide a possible method for assessing performance in response to severe external fire sources (such as back of house fires for commercial/industrial buildings).

Both tests simulate an external (or internal) fire source located directly against the base of a re-entrant wall corner.

### 7.1.7.1  25 ft (7.6 m) High Corner Test

The test apparatus structure consists of a two column and girt wall frames and a ceiling frame of joists and metal furring strips to which test wall and ceiling assemblies can be mounted. There is no non-combustible substrate such as concrete or

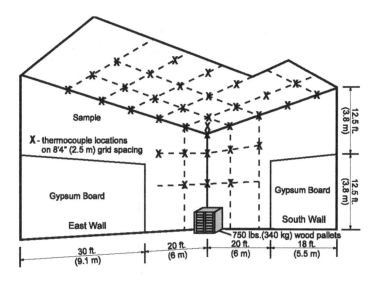

**Fig. 7.9**  25 ft (7.6 m) test apparatus (from ANSI FM 4880-2001R2007 [92])

masonry. The height to the underside of the ceiling frame is 7.54 m. One wall is 15.7 m wide and the other wall is 11.96 m wide. For tests on wall assemblies only, corrugated steel decking is installed to the underside of the ceiling frame. The test wall is installed representative of the end use, which typically involves through bolting of insulated sandwich panels directly to the frame. Test walls are installed to top half (above 3.8 m) extending over the entire width of each wall. Test walls are installed to the bottom half (below 3.8 m) extending only 6 m from the corner on each wall. The remaining sections of the wall are clad with gypsum board.

The fire source is $340 \pm 4.5$ kg crib constructed of 1.065 m oak pallets stacked to a maximum height of 1.5 m and located in the corner 305 mm from each wall. The crib is ignited using 0.24 L of gasoline at the base of the crib. The standard does not state any calibrated heat flux or temperature requirements for the fire source. However it is understood that the maximum heat flux is 100 kW/m² or greater.

Thermocouples are located on the test walls on 2.5 m grid spacing. The test duration is 15 min.

The performance requirement for this test is that the tested assembly shall not result in fire spread to the limits of the test structure as evidenced by flaming or material damage (Fig. 7.9).

### 7.1.7.2  50 ft (15.2 m) High Corner Test

The test apparatus structure consists of two wall frames and a ceiling frame to which test wall and ceiling assemblies can be mounted. There is no non-combustible substrate such as concrete or masonry. The height to the underside of the ceiling frame

**Fig. 7.10** 50 ft (15.2 m) test
apparatus (from ANSI FM
4880-2001R2007 [92])

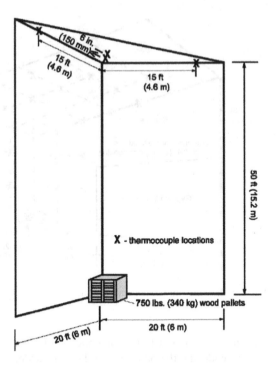

is 15.2 m. Both walls are 6.2 m wide. For tests on wall assemblies only, corrugated steel decking is installed to the underside of the ceiling frame. The test wall is installed representative of the end use, which typically involves through bolting of insulated sandwich panels directly to the frame. Test walls over the entire height and width of the test frame.

The same fire source as for the 25 ft high corner test is used.

Thermocouples are located near the intersection of the top of the walls and the ceiling both at the corner and 4.6 m out from the corner. The test duration is 15 min (Fig. 7.10).

The performance requirements for this test are:

- The tested assembly shall also meet the requirements of the 25 ft corner test
- For acceptance to a maximum height of 50 ft (15.2 m) the tested assembly shall not result in fire spread to the limits of the test structure as evidenced by flaming or material damage.
- For acceptance to an unlimited height the tested assembly shall not result in fire spread to the limits of the test structure or to the intersection of the top of the wall and the ceiling as evidenced by flaming or material damage.

## 7.1.8   Full Scale Façade Testing in China

The test method for fire performance of external wall insulation systems applied to building façades is a test is similar in its dimensions and measurements with the BS8414-1 where the substrate is a masonry wall. This test method for fire-resistant performance of external wall insulation systems applied to building façade is described in GB/T 29416−2012 (Chinese standard [106]).

The test apparatus for the fire performance of building exterior insulation systems consists of L-shaped walls, a combustion chamber, a heat source, a collapse zone and measurement systems (Fig. 7.11) The test apparatus shall be constructed in a space large enough to meet the construction and installation requirements of the wall specimen and test operation, and ensure availability of air for combustion. The test apparatus shall be durable during the test.

Wall: The wall consists of a main wall and a side (wing) wall (L-shaped), using a dry density of not less than 600 kg/m$^3$ of vertical autoclaved aerated concrete block masonry. The height of the main wall and the side wall should be larger than 9000 mm, and the thickness larger than 300 mm. The width of the main wall is larger than 2600 mm, and the side wall larger than 1500 mm. The test wall should be processed by M10 ordinary plaster dry-mixed mortar in the surface, according to GB/T 25181 Chinese standard with a thickness of $(10 \pm 1)$ mm.

Burning room (Combustion chamber): The burning room is located in the bottom of the main wall, with its outer edge flush to the main wall. The opening size has a height of $(2000 \pm 100)$ mm and a width of $(2000 \pm 100)$ mm. The dimensions of the burning room are: height $(2300 \pm 50)$ mm, width $(2000 \pm 50)$ mm, depth $(1050 \pm 50)$ mm. The distance of the opening and side to the wall edge is $(250 \pm 10)$ mm. The top of the opening should be covered by fire resistant materials.

Heat source: A wood crib or a gas burner can be used as heat sources. The heat source should generate the flame to be ejected from the opening and spread upwards.

Collapse area: The collapse area is in the cross angle of the main wall and side wall of length 2450 mm and width 1200 mm, with mark on the ground (Fig. 7.12).

Measurements: The measuring devices include thermocouples, a data collection system, Video recorders, timer and Anemometer.

The thermocouples are divided into internal and external groups. External thermocouple measurement points should extend beyond the outer surface of the insulation system for $(50 \pm 5)$ mm, with tolerance of $\pm 10$ mm. Internal thermocouple measurement points should be arranged in the centre of the thickness in the insulation layer. When the insulation thickness is smaller than 10 mm, there is no need to install the thermocouples. If the system contains a cavity, the internal thermocouple measurement points should also be arranged in the thickness of each centre of the cavity. Temperature measurement point position tolerance of $\pm 10$ mm.

The location of thermocouples in horizontal line is:

- On the surface of the main wall, thermocouples are located in the centreline, and both sides with distance 500, 1000 mm from centreline (Fig. 7.11).

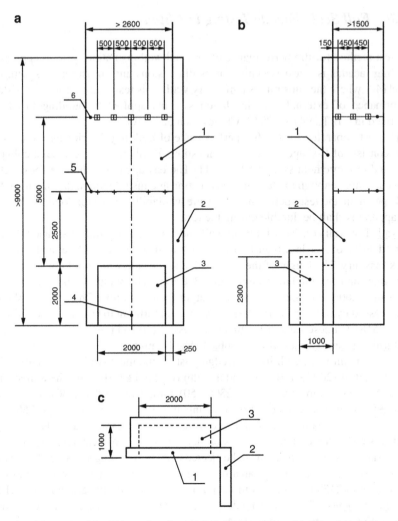

**Fig. 7.11** Schematic of the Chinese façade test [106]. (**a**) Front view. (**b**) Side view. (**c**) Top view. *1* main wall, *2* side wall, *3* burning room, *4* center line of burning room, *5* horizontal line 1, *6* horizontal line 2, +—thermocouple in horizontal line 1 (external temperature); ⊞—thermocouple in horizontal line 3 (internal and external temperature)

**Fig. 7.12** Collapse area between the walls [106]

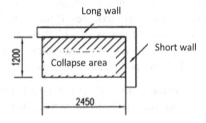

- On the surface of the side wall, thermocouples are located in 150, 600 and 1050 mm from the main wall, three points along the horizontal line 1 and line 2 (Fig. 7.11).

  Internal thermocouples in horizontal line 2

- inside the main wall insulation system, the thermocouples are installed in the centerline of the opening and 500, 1000 mm from the centreline at both sides (Fig. 7.11).
- inside the side wall insulation system, the thermocouples are installed 150, 600 and 1050 mm from the main wall (Fig. 7.11).

Test Sample: The installation of the test sample cannot block the opening of the burning room having thickness not more than 200 mm on the main wall of the apparatus and the width of the sample should not be less than 2400 mm, with one side attached to the surface of the side wall. The location from the top of the opening should be over 6000 mm on side wall of the apparatus, the width of sample should not be less than 1200 mm, with one side attached to the surface of the main wall. The location from the top of the opening should be over 6000 mm.

## 7.1.9   Full-Scale Façade Test in Japan

A new facade test is proposed in Japan [74] as illustrated below in Fig. 7.13 however this test is not applied as part of current building regulations. A gaseous burner of varying heat release rate can produce facade flames of varying flame heights and intensities.

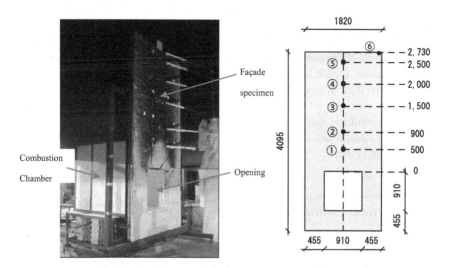

**Fig. 7.13**  Proposed Facade Test in Japan including instrumentation [74]

### 7.1.10   Other Full Scale Façade Fire Tests

The full scale tests described above are the main test of their type applied around the world. However other full scale tests exist in some countries. The following other tests are described in a paper by Smolka et al. [107].

- *LEPIR II* [108]—French façade fire test used for development purposes. No approval criteria are linked to this method in French regulations
- *MSZ 14800-6* [109]: A real-sized approval test applied in Hungary, originated from LEPIR. A revision is on-going; adding a lateral wing to the flat wall is under consideration.
- *Önorm B 3800-5* [110]: A variation of the draft DIN method that can be used for product approvals in Austria.
- *GOST 3125* [111]: Similar to the MSZ14800-6 method. Used in Russia and associated countries

## 7.2   Intermediate-Scale Façade Fire Test Methods

### 7.2.1   ISO 13785:2002 Part 1: Intermediate Scale Facade Test [11]

The test façade is installed as a re-entrant corner "L" arrangement with a total specimen height of 2.4 m, a rear wall width of 1.2 m and side wall width of 0.6 m. The façade is installed representing the end use with all cavity insulation, air gaps and fixings include. The fire source is a linear propane burner 1.2 m×0.1 m in area which is located 0.25 m below the bottom edge of rear wall. The burner has a constant 100 kW output which is sufficient to achieve direct flame impingement on the bottom 200 mm of the rear wall façade. Temperatures are measured vertical intervals of 0.5 m on the centre of both façade wall surfaces. Heat Flux is measured at the top of the rear façade wall. Fire spread is observed (Fig. 7.14).

The test standard does not provide any acceptance criteria and does not provide details of any correlation between performance in the Part 1 test and the Part 2 test. The ignition source is significantly smaller than for full scale tests. How this test does provide a useful and less expensive method for quickly screening and comparing alternative systems.

### 7.2.2   Vertical Channel Test

The vertical channel test was originally developed by NRC to provide a cost effective intermediate test than the full scale CAN/ULC S134 test method. The intent was to achieve the same exposure conditions as the full-scale test. A series of tests

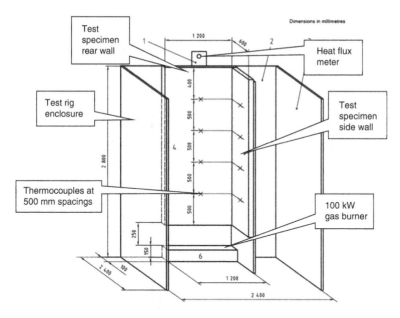

**Fig. 7.14** ISO 13785 Part 1 test rig [11]

carried out by NRC demonstrated that the vertical channel test correlated well with the full scale test [24]. The test method was published as an ASTM Draft proposed rest method [112].

The ASTM Vertical channel test is conducted on a single wall with façade, cladding or exterior wall system that is 800 mm wide and 7.32 m high. The specimen is installed representative of the end use including all insulation, cavity air gaps and fixing details. The test specimen wall is located at the rear of a channel formed by non-combustible 500 mm wide vertical projections one each side of the specimen wall. The purpose of this channel is to enhance the fire exposure conditions to the reduced width specimen produced by a reduced fire source size.

The fire source is intended to simulate flame spread from a compartment fire via a window opening. The fire source is two propane gas burners located in a combustion chamber 1.9 m high×1.5 m deep×0.8 m wide located at the base of the test wall. The combustion chamber has two openings across the widths of the chamber at the front in line with the front of the test wall. The lower opening is 440 mm high and is an air inlet. The top opening is 630 mm high and is a flame outlet. The burners are controlled to achieve a heat flux of $50 \pm 5$ kW/m$^2$ at 0.5 m above the opening and $27 \pm 3$ kW/m$^2$ at 1.5 m above the opening averaged over a 20 min period of steady burner output. This is typically achieved with a propane supply of 25 g/s (1.16 MW). During the test heat flux is measured at the front face of the test wall 3.5 m above the opening and temperatures are measure at the front surface and at each intermediate layer at intervals of 1 m starting at 1.5 m above the opening. The test duration is 20 min (Fig. 7.15).

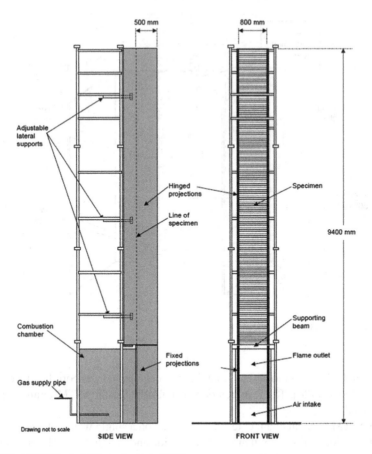

**Fig. 7.15** ASTM Vertical Channel test rig [112]

The test acceptance criteria are:

- Flame does not spread more than 5 m above the bottom of the specimen
- Heat flux 3.5 m above the opening does not exceed 35 kW/m$^2$

In 2005, BRANZ developed a modified version of the vertical channel test and undertook a series of tests investigating the use of the cone calorimeter test as a pre-screening test for external combustible wall linings [113]. The main changes to the vertical channel test by BRANZ were:

- Reduction of the specimen wall height to 5 m
- Some modification to gas supply rate and combustion chamber ventilation conditions to better match the full scale test exposure.

## 7.2.3   FM Parallel Panel Test [114, 115]

FM Global has developed a parallel panel test as an intermediate scale test to predict results for the 25 and 50 ft corner tests. The parallel panel test apparatus consists of two parallel panels, each 4.9 m high by 1.1 m wide, separated by 0.5 m. A sand burner, 1.1 m by 0.5 m by 0.3 m high, is located at the bottom of the panels. The total heat release rate from the burning panels during the test is measured by a 5 MW capacity oxygen consumption calorimetry exhaust hood. The burner exposure is controlled to 360 kW to provide a maximum heat flux to the panels of 100 kW/m². This corresponds to the maximum heat flux measured at the panels at the top of the crib in the 25 ft corner test (Fig. 7.16).

A measured HRR of 1100 kW in the parallel panel test was found to represent fire spread to the top of the panels and this criterion is used in additional to visual observation of fire spread which is often difficult due to smoke production.

It was concluded that fire will not propagate to the end of the test array in the 25-ft corner test with combustible wall panels and a non-combustible ceiling if the HRR in the parallel panel test is <1100 kW; fire will not reach the top of the test array in the 50-ft corner test if the HRR in the parallel panel test is less than 830 kW; fire propagation will not reach the ends of the horizontal ceiling in the 25-ft corner test with both combustible wall and ceiling panels if the HRR in the parallel panel test is <830 kW.

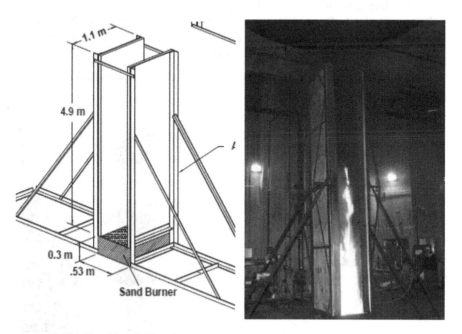

**Fig. 7.16**   FM Global Parallel Panel Test [115]

## 7.3   Room Corner Test Methods

A range of standard room corner test methods exist around the world. These tests simulate the scenario of an interior localised fire occurring in one corner of a room with a ventilation opening (typically a door) and they evaluate the propensity for fire spread on interior wall and ceiling linings resulting in flashover. In some tests the wall and ceiling linings are fixed to a non-combustible lined test room substrate and in others materials such as insulated sandwich panels are constructed as a self supporting, free standing test room so that structural integrity and collapse can also be evaluated under fire conditions. (Opening up of joints in such systems can significantly influence fire growth) (Fig. 7.17).

Room corner tests certainly are not intended to assess fire performance of external walls and facades. However, good performance of a material in a room corner test is sometimes used (particularly by fire engineers justifying an alternative solution) to indicate acceptable performance of the same material as exterior wall assembly. Whilst this may give some degree of confidence in performance the following issues must be considered:

- The ignition source HRR for a room corner test simulates a localised pre-flashover fire and is significantly lower than the worst case scenario identified for exterior wall assemblies, being a post flashover fire with flames ejecting from an opening
- The orientation and exposure of materials in the room fire test can be significantly different to an exterior wall system.
- Room corner tests do not expose or test the edge treatment/design of the window opening and therefore the propensity for fires to spread into the internal cavity of the wall system via this opening is not tested.

The following table provides a brief summary of the various different room corner test methods (Table 7.1)

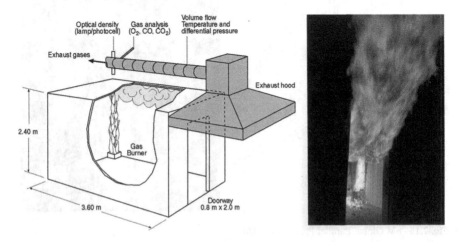

**Fig. 7.17** ISO 9705 room corner test layout and resulting flashover (CSIRO)

**Table 7.1** Summary of room corner test methods

| Test method | Fixed linings inside non combustible test room or free standing room test | Room dimensions | Ventilation opening | Ignition source | Measurements |
|---|---|---|---|---|---|
| ISO 9705 [116] | Fixed | 2.4 m wide × 2.4 m high × 3.6 m long | 0.8 m × 2.0 m doorway | Gas burner with output of 100 kW for 0–10 min and 300 kW for 10–20 min | HRR |
| | | | | | Smoke optical density |
| | | | | | Temperatures at ceiling level and opening |
| | | | | | Heat flux at floor level |
| NFPA 286 [117] | Fixed | 2.44 m wide × 2.44 m high × 3.66 m long | 0.78 m × 2.02 m doorway | Gas burner with output of 40 kW for 0–5 min and 160 kW for 5–15 min | HRR |
| | | | | | Smoke optical density |
| | | | | | Temperatures at ceiling level and opening |
| | | | | | Heat flux at floor level |
| UBC 26-3 [118] | Fixed | Interior dimensions 2.44 m wide × 2.44 m high × 3.66 m long | 0.78 m × 2.13 m doorway | Douglas Fir timber crib 13.6 kg, 381 mm square base area, each stick 38 mm square. Five sticks per tier | Temperatures at ceiling level and opening |
| | | | | | Internal panel temperatures |
| | | | | | Visual observation of fire spread, flashover damage and smoke |

(continued)

**Table 7.1** (continued)

| Test method | Fixed linings inside non combustible test room or free standing room test | Room dimensions | Ventilation opening | Ignition source | Measurements |
|---|---|---|---|---|---|
| ISO 13784 Part 1 [119] | Free standing | 2.4 m wide × 2.4 m high × 3.6 m long | 0.8 m × 2.0 m doorway | Gas burner with output of 100 kW for 0–10 min and 300 kW for 10–20 min | HRR<br>Smoke optical density<br>Temperatures at ceiling level and opening<br>Heat flux at floor level<br>Internal panel temperatures |
| ISO 13784 Part 2 [120] | Free standing | 4.8 m wide × 4.0 m high × 4.8 m long | 4.8 m × 2.8 m doorway | Gas burner with output of 100 kW for 0–5 min and 300 kW for 5–10 min and 600 kW for 10–15 min | Internal and surface panel temperatures<br>Visual observation of fire spread, flashover and damage |
| LPS 1181 Part 1 and Part 2 [121, 122] | Free standing | Large free standing room fire test (10 m L × 4.5 m W × 3 m H). Applies timber crib | 2.25 × 4.5 m W opening | Redwood/Scots Pine timber crib. Seventy sticks of 50 mm × 25 mm × 750 mm | Temperatures at ceiling level and opening<br>Internal panel temperatures<br>Visual observation of fire spread, flashover and damage |

## 7.4   Small-Scale Test Methods

### 7.4.1   Combustability Tests

Combustibility tests are essentially used to determine if materials are combustible or non-combustible. Various standard test methods exist around the world including (ISO 1182, BS 476 part 4, ASTM E136, ASTM E2652, AS 1530.1) [123–127] however they are all fairly similar.

Small specimens are exposed to high temperatures of typically 750 or 835 °C within a small conical tube furnace. Criteria for non-combustibility are typically.

- No sustained flaming (typically >5 s)
- Mean furnace temperature rise must not typically exceed 50 °C
- Mean specimen surface temperature must not typically exceed 50 °C
- Criteria for limited specimen mass loss may also be applied.

Many building codes around the world deem materials such as gypsum plaster to be non-combustible as they don't necessarily meet the above test criteria for items such as mass loss.

External wall assemblies constructed entirely of non-combustible materials do not generally pose any hazard relating to fire spread.

### 7.4.2   Cone Calorimeter

The cone calorimeter [128] is a small-scale oxygen consumption calorimeter. Specimens, 100 mm square are supported horizontally on a load cell and exposed to a set external radiant heat flux in ambient air conditions. The radiant heat source is a conically shaped radiator that can be set to impose any heat flux in the range 0–100 kW/m² on the specimen surface. Ignition is promoted using a spark igniter. Combustion gases are extracted in an exhaust duct where instrumentation measures exhaust gas flow, temperature, $O_2$, CO and $CO_2$ concentrations and smoke optical density. From these measurements quantities such as heat release rate, mass loss rate, effective heat of combustion and smoke production can be calculated. Time to ignition at set heat flux exposures is determined by observation. The cone calorimeter apparatus and procedure are described in ISO 5660, AS/NZS 3837 and ASTM E 1354 (Fig. 7.18) [129–131].

The cone calorimeter attempts to measure fundamental flammability properties of materials that are required to predict material behaviour in real fires. Much research has been focused on predicting real fire behaviour based on cone calorimeter results, however the ability to make such predictions remains very limited. Some reasons for this are;

- The cone calorimeter method measures properties under set conditions which affect the properties attempting to be measured,

**Fig. 7.18**  Cone Calorimeter (CSIRO)

- The cone calorimeter does not directly measure all fundamental properties that may be required such as heat of volatilisation, heat capacity and thermal conductivity, and
- The theoretical link between fundamental properties and real fire behaviour is complex and not well developed.

For materials which are complex composites with protective external layers that have a low combustibility the cone calorimeter often fails to predict the true hazard of the combustible core material which may become exposed in a full-scale fire due to fail of joints etc. The cone calorimeter also has limitation when testing materials with reflective surfaces due to the large amount of heat reflection.

The cone calorimeter is a very complex apparatus requiring more maintenance and calibration than other small-scale fire apparatus. Erroneous data can easily be generated if the operator does not have a high level of competency.

Despite these limitations the cone calorimeter is still one of the most useful tools for determining flammability properties for materials.

### 7.4.3  Euroclass Tests

The Euroclass system for characterising reaction to fire behaviour of construction products is applied throughout most of Europe and is specified in EN 13501-1 [83]. The Euroclass system was designed for controlling flammability of internal materials and does not specifically address exterior wall systems. However due to a lack of any uniform approach throughout Europe to control exterior wall systems via harmonised requirements for either small or large scale testing, individual European

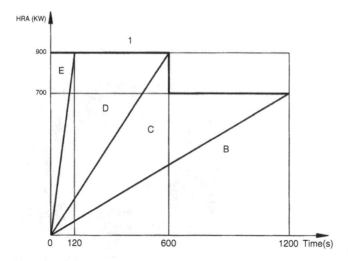

**Fig. 7.19** Relationship between Euroclasses and ISO 9705 room corner test time to flashover [83]. *1* flashover, *B* class B/A2, *C* no flashover for 100 kW but flashover, *D* flashover after more than 2 min for 100 kW ignition source, *E* flashover before 2 min for 100 kW ignition source. *Note* HRR from the specimen excludes the burner

countries have resorted to either relying on Euroclasses or national large scale façade tests for control of exterior wall systems.

It is often applied to exterior wall systems.

For non flooring materials the Euroclass system applies a range of small scale tests and is intended to classify materials in terms of contribution to fire development for a scenario of a fire starting in a small room by a single burning object. As follows:

- Class A1 products are essentially non-combustible and will not contribute to fire growth nor to the fully developed fire
- Class A2 products have a very low combustibility and will not significantly contribute to the fire growth and fuel load in a fully developed fire
- Class B products are combustible, will not lead to a flashover situation but will contribute to a fully developed fire
- Class C–E products may lead to flashover at the reference scenario test times shown in Fig. 7.19

For non flooring materials the four following tests are applied to determine the classification

**EN ISO 1182 Non Combustibility [123]**

See Sect. 7.4.1.

**EN ISO 1716, Gross Calorific Value [132]**

This is an Oxygen Bomb Calorimeter test where a specified mass of material is burnt under standardised conditions within a confined volume combustion

**Fig. 7.20** SBI test [134]

chamber with high oxygen concentration. The Gross calorific potential (heat of combustion) is calculated based on the measured temperature rise of the combustion chamber taking into account heat loss.

### EN 13823 Single Burning Item (SBI) Test [133]

The SBI test is an intermediate scale corner test conducted under an exhaust hood fitted with oxygen consumption calorimetry equipment and smoke meters (typically inside a test room with controlled makeup ventilation). Heat release rate (kW), total heat release (MJ) and smoke production rate ($m^2/s$) are measured. Flame spread and burning droplets are observed visually. The specimen is installed in a corner with a 1 m wide × 1.5 m high long wing and a 0.49 m × 1.5 m high short wing. A 30 kW gas burner is located in the corner and the total test time is 21 min (Fig. 7.20).

### EN ISO 11925-2 Small Flame Test [135]

- The specimens are ignited with a 20 mm high propane gas flame. The flame is impinged on the bottom edge of the specimen (edge exposure) or 40 mm above the bottom edge (surface exposure) or both. The specimen is exposed to flame for 15 or 30 s.
- For each test specimen it is recorded whether an ignition occurs (flaming longer than 3 s), whether the flame tip reaches 150 mm above the flame application point and the time at which this occurs. The occurrence of burning droplets/particles is also observed.
- For each exposure condition a minimum of six specimens (250 mm × 90 mm) of the product shall be tested, three cut lengthwise and three crosswise

Materials are classified based on the above tests as shown in the following table. Only Classes A1-B are shown as lower classes are generally not applied to exterior wall assemblies.

## 7.4.4   British Classification Tests

In addition to the non-combustibility test the UK Approved Document B applies the following British small-scale tests to exterior walls (Alternatively Euroclass tests can be applied).

### 7.4.4.1   BS 476 Part 6 [81]

This Fire propagation test was developed primarily for interior wall linings. The result is given as a fire propagation index. The test specimens measure 225 mm square and can be up to 50 mm thick. The apparatus comprises a combustion chamber attached to a chimney and cowl (with thermocouples). The chamber is heated using electrical elements and a gas burner tube is applied to the bottom of the test specimen. The test specimens are subjected to a prescribed heating regime for a duration of 20 min and the index obtained is derived from the flue gas temperature compared to that obtained for a non-combustible material.

### 7.4.4.2   BS 476 Part 7 [136]

This surface spread of flame test is used to determine the tendency of materials to support lateral spread of flame. The test specimen is rectangular, 925 mm long × 280 mm wide with thickness up to 50 mm. The vertical specimen is mounted perpendicular to a large 900 mm square gas-fired radiant panel. The radiant heat flax along the specimen decreases from 30 kW/m$^2$ at the near end to 5 kW/m$^2$ at the far end. Depending on the extent of lateral flame spread along the specimen, the product is classified as Class 1, 2, 3 or 4 with Class 1 representing the best performance.

### 7.4.4.3   BS 476 Part 11 [82]

This test is very similar to the BS 476 part 4 non-combustibility test. Small samples are exposed to 750 °C in a small tube furnace and the occurrence of any flaming, specimen surface temperature, furnace temperature and specimen mass loss at end of test are measured. UK Approved document B uses this test to classify materials as having limited combustibility.

### 7.4.5   US Building Code Tests

#### 7.4.5.1   NFPA 268: Determining Ignitability of Exterior Wall Assemblies Using a Radiant Heat Energy Source [137]

This test evaluates the propensity for ignition of an exterior wall assembly when exposed to a radiant heat flux of 12.5 kW/m$^2$ and a pilot ignition source over a 20 min test period. The test specimen must be 1.22 m wide×2.44 m high. The gas fired radiant panel is 0.91 m×0.91 m. The radiant panel is stationary and the specimen is mounted on a trolley. The radiant heat flux exposure is controlled by the separation distance. This test only assesses risk of ignition from an external radiant heat source. It does not assess risk of ignition or flame spread from direct flame exposure (Fig. 7.21).

#### 7.4.5.2   ASTM E 84, UL 723, NFPA 255: Steiner Tunnel Test [89, 90, 138]

This test was originally developed for interior wall and ceiling linings and measures both flame spread and smoke production. The test is conducted inside a non combustible horizontal tunnel/box that is 7.3 m long×0.056 m wide×0.305 m high. The specimen is mounted to the ceiling of the tunnel. Gas burners at one end of the tunnel provide a heat output of 89 kW and air and combustion products are drawn through the tunnel in the direction of fire spread at a controlled velocity of 73 m/min. The test duration is 10 min. Flame spread is measured by observation and smoke optical density is measured by an obscuration meter located in the exhaust duct. Results are expressed in terms of a flame spread index and a smoke developed index. Both indices are based on arbitrary scales where cement board has a value of 0 and red oak has a value of 100.

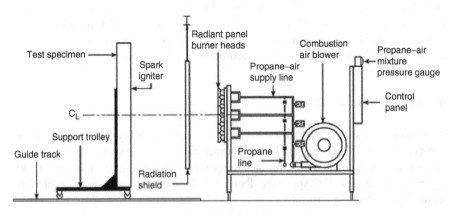

**Fig. 7.21** NFPA 268 test side view (from NFPA 268 [137])

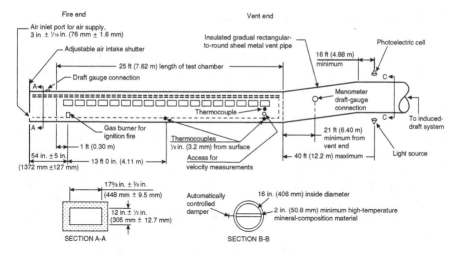

**Fig. 7.22** Steiner Tunnel Test (from NFPA255 [138])

These indices cannot be easily used as basic fire engineering properties or correlated to performance in an exterior wall end use. This test does not properly asses thermoplastic materials which may tend to melt away from the assembly rather than spread flame in the horizontally prone test orientation (Fig. 7.22).

### 7.4.5.3 NFPA 259: Potential Heat of Building Products [139]

This test uses an oxygen bomb calorimeter to determine the heat of combustion for a material. It also specifies placing the same material in a muffle furnace at 750 °C for 2 h and then testing the residue in a bomb calorimeter to determine the potential heat of the residue.

### 7.4.5.4 ASTM D 1929 Standard Test Method for Determining Ignition Temperature of Plastics [91]

This test exposes small pellets of plastic materials to a controlled flow rate of heated air inside a tube furnace. This test measures the two following properties;

- Flash-Ignition Temperature—the lowest initial exposure air temperature at which the combustible gas evolved from the specimen can be ignited by a small external pilot flame.
- Spontaneous-ignition (Self-ignition) temperature—The lowest initial exposure air temperature at which unpiloted ignition of the specimen occurs indicated by an explosion, flame or sustained glow.

### 7.4.6   Small Flame Screening Tests

Small flame tests have been used and misused to test the flammability of materials since the 1930s. During the 1950s and 1960s there was an increased reliance on small flame tests but in recent years this reliance has decreased as new test methods that produce more useful measurements have been introduced [140]. Small flame tests have originated from a need to perform quick and cheap screening tests (such as holding a match to a material to see if it burns) Some methods have become overly complex given these origins. These methods assess the ease of ignition and the ability to sustain flaming under set laboratory conditions but do not provide useful data that can be used to predict fire behaviour for real fire scenarios. They can only be used for screening. Dripping of materials can unseat and extinguish flaming in these tests producing a good test result however in real fire scenarios the material may be orientated or restrained so that it either forms a molten pool or drips onto other combustible materials which may increase hazard of flame spread.

ASTM D 635 [141] is an example of one small flame test which is used in the US IBC relating to exterior wall assembly including plastic panels and metal composite materials. This tests specimens 125 mm long × 13 mm wide in the horizontal position. A Bunsen burner flame is applied for a specified time and time to flame extinguishment, burn distance, linear burning distance and occurrence of flaming droplets are recorded. Other similar small flaming tests that may test in either the horizontal or the vertical position include UL94, IEC 60707, IEC 60695-11-10, IEC 60695-11-20, ISO 9772 and ISO 9773, and EN ISO 11925-2.

## 7.5   Fire Resistance Tests for Curtain Walls and Perimeter Fire Barriers

### 7.5.1   ASTM E2307-10: Fire Resistance of Perimeter Fire Barrier Systems [142]

A perimeter fire barrier is the perimeter joint between the external wall assembly and the floor assembly designed to provide a barrier to floor to floor fire and smoke spread.

ASTM E2307-10 applies the same full scale façade test apparatus as described for NFPA 285. The test rig is a two storey steel framed structure with an open fronted test room on each storey constructed of concrete slabs and walls. Each test room has internal dimensions of approximately 3 m wide × 3 m deep × 2 m high. The bottom test room serves as the fire enclosure and the top test room simulates an enclosure on the level above with no window.

The test determines the ability of the perimeter joint system in combination with the external wall assembly and floor system to maintain an acceptable fire barrier. The perimeter joint tested must be at least 4 m long and is located at the intersection of the external wall with the top level floor. The joint width must be the maximum to be used in end use. The wall system must have a window opening at the lower floor as per NFPA 285 and the wall system may be selected to reflect an end use application (e.g. a particular curtain wall construction). The floor system may also be selected to represent a particular end use system.

Prior to undertaking the fire resistance test the perimeter joint can be exposed to various different movement cycling conditions dependant on the end use movement expected between the wall system and floor system. The same perimeter joint is then exposed to the fire resistance test.

During the first 30 min of the test the window burner and room burners are controlled to achieve the same exposure as required for NFPA 285. From 30 min onwards the window burner is maintained at the 30 min output (nominally 400 kW) and the room burner is controlled so that the average measured temperature at the underside of the floor slab is controlled to follow the ASTM E119 standard temperature time curve (Fig. 7.23).

The fire resistance result of the perimeter barrier is expressed in terms of a "T" rating and an "F" Rating.

The "T" Rating is the time at which one of the following failure criteria occur:

- The temperature rise of any surface thermocouple on the unexposed side of the perimeter barrier (or adjacent supporting structure) exceeds 181 °C above the initial temperature, or
- For maximum joint widths greater than 102 mm, the maximum temperature rise of all unexposed surface thermocouples exceeds 181 °C above the initial temperature.

The "F" rating is the time at which one of the following failure criteria occur:

- Observed flame penetration through or around the perimeter joint, or
- Passage of flames or hot gases sufficient to ignite a cotton pad applied to the unexposed side of the perimeter joint.

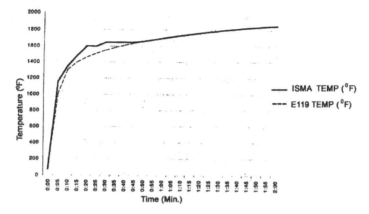

**Fig. 7.23** Typical ASTM E2307 exposure temperature vs. ASTM E119 standard time temperature curve (from ASTM E2307 [142])

This test method has the advantage that it simulates a room fire with flames eject-ing from the window so that the wall assembly is heated from both sides which may influence any movement of the wall during the test. It tests the performance of the joint system in combination with the wall and floor system. However the results for a specific combination of systems may not be applicable if on part (such as the wall system) is significantly changed.

It may be possible to combine assessment of the fire resistance of the perimeter joint with assessment of fire spread on the exterior wall into a single test by combin-ing the requirements of NFPA 285 and ASTM E2307-10.

### 7.5.2   EN 1364 Part 3 Fire Resistance Test for Non Loadbearing Curtain Walls [143]

EN 1364 Part 3 evaluates the fire resistance performance of complete curtain wall systems. The test applies a standard 3 m×3 m vertical fire resistance furnace and a standard time vs. temperature curve.

The curtain wall must be constructed to be representative of the end use con-struction. Restraint of the curtain wall (typically to the top and bottom edges of the furnace) must represent end use. The test may be conducted either as an interior fire exposure or an exterior fire exposure by orientating the wall correctly. For interior fire exposure tests vertical and or horizontal gap seals representative of the end use design may be included between the curtain wall and the furnace along the side and top edges.

The failure criteria for this fire resistance test are the standard failure criteria relating to integrity and insulation as specified in EN1364 part 1 and other similar fire resistance tests around the world. Additionally there is a requirement to mea-sure deflection of the curtain wall at its centre point as well as 50 mm from the free edge.

EN 1364 Part 4 [144] provides a method for testing individual components of curtain wall systems such as edge seals, mullions etc. rather than the entire system.

# Chapter 8
# Recommended Fire Scenarios and Testing Approach for Phase II

Phase I of this study seeks to collect data about combustible facade systems, review existing research in this area, examine statistics on façade fires, list incidents of facade fires, describe the mechanisms and dynamics of fire spread and review existing test methods and performance criteria. This section presents the main conclusions of Phase I and recommends various options for further testing and research for Phase II.

As the Fire Research Foundations objectives for Phase II have not been specifically stated at this stage, we suggest a range of options for further experimental work for consideration.

## 8.1 Recommended Fire Scenarios

The consensus from existing research is that full-scale façade tests should simulate an internal post flashover fire with flames ejecting from windows as this is considered to be more severe than an external fire source, and therefore suitable cover this alternative fire scenario.

Based on the present review for large scale façade existing test methods we make the following comments:

- All of the façade tests reviewed simulate an internal post flashover fire with flames ejecting from windows.
- However it is possible for the severity external fires at ground level on fuel loads such as back of house storage areas and large vehicle fires to equal or exceed internal post flashover fires. The impact of exterior fire sources can be even more severe if they occur hard against re-entrant exterior wall corners. Although most full-scale façade tests simulate an internal post flashover fire, these tests may also set a suitable level of performance with regards to a limited external fire severity.

© Fire Protection Research Foundation 2015
N. White, M. Delichatsios, *Fire Hazards of Exterior Wall Assemblies Containing Combustible Components*, SpringerBriefs in Fire,
DOI 10.1007/978-1-4939-2898-9_8

- Dimensions and physical arrangement of facade tests vary. As an example, some large-scale tests involve external corner walls 8 m high (UK) or 5.7 m high (Germany and ISO) and 2.4 and 1.3 m wide. Some tests such as NFPA 285 and SP105 involve a single wall rather than a corner
- There are significant differences in the source fire simulating a fire in the room of origin. Wood cribs, liquid pool fires and gas burners are being used to generate maximum measured heat fluxes on the façade in the range of 20–90 kW/m². The speed of ejection of the hot gases changes the impact on the specimen. Flames can project outward from the wall when at velocity or adhere to the wall and test specimen when the plume is lazy (lower velocity flow). Selection of the fire should be based on control of the impact of the test fire on the sample.
- Test durations, measurements and acceptance criteria vary.
- The degree to which suitability of fixing systems and fire spread through joints, voids and window assemblies of a multifunctional façade assembly are tested varies.
- Whilst large-scale facade tests do not characterise and assess the individual elements of the facades, these tests do provide useful information on end use fire spread performance which can be used for validation of intermediate tests, small scale tests and fire spread modelling.
- The case studies of fire incidents indicate more extensive and rapid flame spread occurs where there is a wing wall (re-entrant corner), channel or part channel formed by balconies.

Fire exposure to external walls from real post flashover fire scenarios is dependent on:

- Room geometry
- Room fuel load
- Opening size and ventilation conditions.

As the above factors can vary significantly in reality from building to building, it is reasonable that the various full scale façade tests with different wall geometry and fire exposure severity are valid for fire scenarios within a range of these factors.

Therefore development of a new full-scale test to simulate a specific fire scenario is not recommended at this stage. Instead further research to validate the existing full-scale and small scale tests and also to develop a more affordable and dependable intermediate-scale test are recommended.

Option 4 does suggest investigation of the effect of changing the geometry of an existing full-scale test to a large vertical "U" shaped channel.

Additionally work to develop façade fire spread models is recommended.

## 8.2   Recommendations for Further Test Based Research for Phase II

The following are options for consideration. Multiple options may be adopted. It is understood that the cost of full-scale testing may be prohibitive for some options.

## 8.2.1   Option 1: Existing Full-Scale Facade Test Round Robin

The range of geometry, fire exposure and acceptance criteria for the various existing full scale façade tests may potentially result in variation of test results and acceptance of materials in different countries.

To examine this further and to gather full-scale test data to support other research options a full-scale façade test round robin could be conducted between labs around the world which are currently operating the different tests. The following are preliminary recommendations for such a round robin.

- Prior to conducting the round robin it should be determine if a similar project has already been undertaken. A test based comparison of BS 8414-1 &-2, Draft DIN 4102-20 and ISO 13785-1 & -2 has recently been reported by BRE [103]. The author is aware of an ISO round robin for ISO 13785.2 proposed at the time of report finalisation.
- To provide a basis for comparison, each test should be fitted with heat flux gauges and thermocouples at consistent heights which are in addition to the standard instrumentation.
- The round robin should be conducted applying the same exterior wall system to each test. Prior to conducting the round robin existing large scale test data should be reviewed to select a suitable test wall system which is most likely to discriminate any differences between the tests. Ideally at least three different wall systems ranging in performance from good to poor would be applied to each full-scale test method to obtain a useful spread of test data. However this large number of tests may be prohibitively expensive.
- Measurement of heat fluxes on an inert façade with the above additional instrumentation should also be performed to characterise the fire exposure in detail.

The expected benefits of this option would be

- Increased understanding of the relative performance of the different test methods and identification of comparative weaknesses in any test methods.
- Basis for accepting a system tested to one particular test method in other countries which reference different test methods.
- Would provide full-scale test data to support other research options suggested.

## 8.2.2   Option 2: Development and Validation of Intermediate Scale Facade Test

Development and validation of a suitable intermediate scale façade test prior to testing in full-scale. This option could involve the following

- Reviewing the current intermediate scale façade tests including ISO 13785 Part 2, the Vertical channel test and the current intermediate façade test proposed in Japan to select an appropriate intermediate test for further development and validation. The review should include any existing test data comparing intermediate scale tests to full-scale tests.

- Identify the particular full-scale test to be applied as the reference for validation.
- Identify any changes to the test apparatus or method required to improve the selected intermediate-scale tests.
- Identify critical aspects of exterior wall construction that must be included in small or intermediate scale testing to ensure poor performance of materials are correctly demonstrated. For example, fire exposures will typically be less severe than for real façade fires, therefore partial exposure of combustible core materials in these test may need to be considered to represent degradation of protective non combustible exterior layers.
- Conduct a series of tests to validate the ability of the intermediate-scale test to predict the outcomes of full-scale tests. Tests should be on systems where full-scale test data is already available either from the Option 1 round robin tests or other prior tests.

### 8.2.3   Option 3: Validation of Small-Scale Test Regulatory Requirements

In countries where exterior wall materials are regulated using small scale tests (or where exemptions from Full-scale tests are permitted based on small-scale test results) it is not apparent that the requirements are based on test based validation. In order to ensure that regulatory requirements for small scale tests are appropriate the following could be undertaken.

- Select the country/countries and test methods of concern.
- Review if any validation of small-scale test regulatory requirements has previously been undertaken.
- Collate any exiting small-scale and full-scale test data on a range of exterior wall systems that can suitably be applied to validate requirements
- Indentify and carry out any further small-scale and full-scale testing that may be required to validate requirements
- Examine test data investigate any correlations and limitation of small scale tests vs. large scale performance and conclude on the suitability of existing regulatory requirements.
- Investigate the possibility of assessing the performance of individual façade components, which in combination with proper fire breaks would give a better assessment of the behaviour of a full-scale façade. For this purpose tests such as the SBI test may be much more appropriate than other small scale tests. Although the SBI test is not an intermediate scale façade test. Although the SBI test is not an intermediate scale façade test, it does represent the vertical flame spread scenario for external wall systems more realistically then other small scale tests, although the 30 kW ignition source is relatively small.

## 8.2.4   Option 4: Investigation of Effect of Vertical "U" Channel on Full-Scale Test

Some of the fire incidents reviewed having very rapid fire spread involved external vertical "U" shaped channels extending over a significant height of the building created by balconies and the like on the exterior. Examples of this were the Mermoz Tower fire in France and the Wooshin Golden Suites fire in South Korea, however both of these fires also involved poor performing metal composite panels.

It is expected that this external profile enhances re-radiation to combustible surfaces and creates a chimney effect increasing vertical ventilation of the fire. It is not known if this external profile would result in poor performance/failure for materials which pass the existing full-scale tests which either have a single wall or an "L" shaped corner.

To investigate this it is suggested that the following testing could be undertaken.

- Select a given full-scale façade test such as NFPA 285 (single wall) or Bs8414 (corner)
- Select a given wall systems for which previous test results (in the normal configuration) exist. Preferably two different wall systems should be tested (one which was a borderline pass and another which performed well)
- Reconfigure the full-scale façade test to include a rear wall (2 m wide suggested) and two wing walls (at least 1.5 m wide suggested)
- Redo tests on selected materials with re-configured geometry.
- Redo tests on selected materials with re-configured part channel geometry.

The purpose of this is to investigate:

- If this geometry has any significant impact on performance of materials which pass the test in the standard geometry
- Are any increased requirements needed for materials that are to be installed in this arrangement in end use?
- Compare of fire spread on materials in the following configurations:
  - Flat wall
  - Wing wall
  - Channel
  - Part channel formed by balconies

It is expected that this may result in a test which is too severe for normal regulation of materials, and is not recommended to replace the existing full-scale test geometries however this may inform regulations on the limitations of combustible materials in a channel configuration. Assessment of this situation and the development of such a test ("U" Shaped façade with side wing walls) may be assisted by the recent work by FireSERT and USTC (China) on facade flame heights with side walls [2].

## 8.2.5 Option 5: Development of Façade Flame Spread Models

Modelling of vertical flame spread is a complex problem, particularly for wall systems which are complex assemblies consisting of multiple different layers of materials, cavities and fire stopping.

Although flame spread models have the potential to provide the link between small-scale and full-scale test performance, the models currently available do not deal well with the above complexities and are not considered valid or reliable enough to eliminate the need for full-scale testing. Flame spread modelling is definitely in the realm of research rather than practical application for regulation at this point in time.

However continued research on developing and validating flame spread models is required to move beyond these current limitations. This option could include

- Review of existing façade fire spread models including CFD and empirical models to identify models which have the greatest potential for successful prediction.
- Measure key flammability properties of the combustible facade components in small scale tests including the cone calorimeter determine required model input data. This could be obtained via Option 3 above
- Obtain intermediate scale façade test results and apply models to firstly predict intermediate façade results. This could be obtained through Option 2 above.
- Obtain full scale façade test results and apply models to predict full-scale façade results. This could be obtained through Option 1 above.
- It is noted that existing models are likely to only apply to simple homogenous materials (not complex systems)
- if testing of models over a wider range of materials is required then this could be more easily and cheaply tested using a suitable intermediate scale test rather than full-scale testing

An alternative or parallel performance based approach, which can also be used for risk analysis, is also proposed:

1. Assess and Measure key flammability properties of the combustible facade components in small (cone calorimeter) and intermediate scale experiments (SBI). Based on these tests and analysis, classify, for example, the materials according to European regulations for construction products. Then for regulation, Euroclass B or better may be accepted for individual components.
2. Determine size of fire for the specific enclosure in the built environment based on recent research work.
3. Reproduce this fire size using a gas burner in a test similar, for example, to one proposed and developed in Japan as Option 2.
4. Measure and/or model the heat fluxes of facade flames on an inert façade in the selected test
5. Test the real facade assembly and use the results to assist in establishing regulations
6. If load bearing facade, perform also a fire resistance test with conditions reproducing the heat fluxes in Part 3.

# Chapter 9
# Conclusions

The following conclusions are drawn from Phase I of this project:

- A broad range of different types of combustible exterior wall assemblies are in common use including exterior insulated finish systems (EIFS), metal composite cladding, high pressure laminates and a range of other systems. These exterior wall systems are typically complex assemblies of different material types and layers which may include vertical cavities with or without fire stopping.
- The key initiating fire can be one of two possible types of fires:

  I. Fires external to the building (other burning buildings, external ground fires) or
  II. Fires internal to the building originating in a floor that have resulted in breaking the windows and ejecting flames on the façade

- Key mechanisms of fire spread after initiating event include:

  I. Fire spread to interior of level above via openings such as windows causing secondary interior fires on levels above resulting in level to level fire spread (leap frogging)
  II. Fire spread on the external surface of the façade assembly, if combustible
  III. Flame spread within an interval vertical cavity /air gap or internal insulation layer. This may include possible failure of any fire barriers if present, particularly at the junction of the floor with the external wall.
  IV. Heat flux impacts causing degradation/separation of non-combustible external skin (loss of integrity) resulting on flame spread on internal core
  V. Secondary external fires to lower (ground) levels arising from falling burning debris or downward fire spread.
  VI. Channelling of convective heat and re-radiation between surfaces such as corners or in channels can accelerate flame spread.

© Fire Protection Research Foundation 2015
N. White, M. Delichatsios, *Fire Hazards of Exterior Wall Assemblies
Containing Combustible Components*, SpringerBriefs in Fire,
DOI 10.1007/978-1-4939-2898-9_9

- Statistics relating to exterior wall fires have been reviewed. Statistical data relating to exterior wall fires is very limited and does not capture information such as the type of exterior wall material involved, the extent of fire spread, or the mechanism of fire spread. Exterior wall fires appear to account for somewhere between 1.3 % and 3 % of the total structure fires for all selected property types investigated. However for some individual property types exterior wall fires appear to account for a higher proportion of the structure fires, the highest being 10 % for storage type properties. . This indicates that exterior wall fires are generally low frequency events, particularly compared to fires involving predominantly the interior.
- The percentage of exterior wall fires occurring in buildings with sprinkler systems installed ranges from 15–39 % for the building height groups considered. This indicates that whilst sprinklers may have some positive influence, a significant percentage of external wall fires still occur in sprinkler protected buildings, which may be due to both external fire sources or failure of sprinklers. On this basis it is recommended that controls on flammability of exterior wall assemblies should be the same for sprinkler protected and non-sprinkler protected buildings
- Review of fire incidents around the world indicates that although exterior wall fires are low frequency events, the resulting consequences in terms of extent of fire spread and property loss can be potentially very high. This has particularly been the case for incidents in countries with poor (or no) regulatory controls on combustible exterior walls or where construction has not been accordance with regulatory controls.
- Combustible exterior wall systems may present an increased fire hazard during installation and construction prior to complete finishing and protection of the systems. The 2009 CCTV Tower Fire and 2010 Shanghai fire in China are examples of large fires occurring during construction.
- An overview of existing research related to fire performance of exterior combustible walls is provided. The Fire Code Reform Centre funded a research report on fire performance of exterior claddings [1] provides an excellent overview of the previous research in to the year 2000. Appendix D also provides a list of related research literature for further reading. It follows from this research review that the façade fire safety problem can be divided into four parts:

  o Specification of fire development and the heat flux distribution both inside the enclosure and from the façade flames originating from the fire in the enclosure. This requirement is prerequisite for the following parts.
  o Fire resistance of the façade assembly and façade – floor slab junction including structural failure for non-combustible and combustible façade assemblies
  o Fire spread on the external surface of the façade assembly if combustible due to the flames form the enclosure fire
  o Fire spread and propagation inside the façade insulation, if combustible, due to the enclosure fire

- Regulations vary from country to country. Five aspects of regulation have been identified to influence the risk of fire spread on exterior wall systems. These include reaction to fire of exterior wall systems, fire stopping of cavities and gaps, separation of buildings, separation of openings vertically between stories of fire compartments and sprinkler protection. Of these, the reaction to fire regulation requirements are expected to have the most significant impact on actual fire performance and level of fire risk presented by exterior wall assemblies.
- Some countries including Australia have stringent reaction to fire requirements that the exterior walls must be non-combustible. However in practice combustible systems are applied as fire engineered alternative solutions, however as no full-scale test is used or typically referred to in Australia the basis for alternative solutions is often limited to performance in small scale tests.
- New Zealand primarily applies the cone calorimeter ISO 5660 for regulation of exterior walls. This appears to be the only country to do this.
- Countries such as the USA, UK, and some European countries specify full-scale façade testing but then permit exemptions for specific types of material based on small-scale fire testing. The United Arab Emirates has recently drafted and is applying regulations using full scale façade testing combined with small scale tests in response to a spate of fire incidents involving metal clad materials in 2011–2012
- A range of different full-scale façade tests have been reviewed and are in use around the world. The geometry, fire source, specimen support details, severity of exposure and acceptance criteria varies significantly for different tests. Existing research has identified that exposure to the exterior wall system is generally more severe for an internal post flashover fire with flames ejecting from windows than for an external fire source. For this reason, almost all of the full scale façade fire tests simulate an internal post flashover fire. However it is possible for the severity external fires at ground level on fuel loads such as back of house storage areas and large vehicle fires to equal or exceed internal post flashover fires. The impact of exterior fire sources can be even more severe if they occur hard against re-entrant exterior wall corners. Although most full-scale façade tests simulate an internal post flashover fire, these tests may also set a suitable level of performance with regards to a limited external fire severity.
- Full-scale façade tests with a wing wall are currently the best method available for determining the fire performance of complete assemblies which can be influenced by factors which may not be adequately tested in mid to small scale tests. These factors include the severity of fire exposure, interaction of multiple layers of different types of materials, cavities, fire stopping, thermal expansion, fixings and joints.
- Full-scale tests are usually very expensive.
- Intermediate-scale tests including ISO 13785 Part 2, the Vertical channel test and the Single Burning Item (SBI) test and also a variety of room corner tests are less expensive however they may not correctly predict real-scale fire behaviour for all types of materials due to less severe fire exposures, less expanse of surface material to support fire growth and flame spread, and less incorporation of end use

construction such as joints, fire stopping and fixings etc. Except for the SBI, Intermediate scale tests are currently not used for regulation but may be used for product development.

- A range of different small scale tests exist and are used for regulation in different countries. Small scale tests often are only applied to individual component materials and represent very specific fire exposure conditions. Small scale tests can provide misleading results for materials which are complex composites or assemblies. This is particularly the case where a combustible core material may be covered by a non-combustible or low-combustible material or a highly reflective surface. There is currently no practical method of predicting real scale fire performance from small-scale tests for the broad range of exterior wall systems in common use.
- Small scale tests may provide acceptable benchmarks for individual material components. However further validation against full-scale tests may be required to support this. Small scale tests (in particular the cone calorimeter) can also be useful for doing quality control tests on materials for systems already tested in full-scale or for determining key flammability properties for research and development of fire spread models.
- The test method should include a wing wall and also assess downward fire spread.
- Investigation into the impact channels and part channels formed by balconies have on fire spread should be investigated.

Recommendations and options for further test based research for consideration as Phase II have been provided in Chap. 8.

# Appendix A: US Fire Statistics Tables

The following US fire statistics have been provided by Marty Ahrens and John R. Hall Jr. of the Fire Analysis and Research Division, National Fire Protection Association. The statistics in this analysis are national estimates of building fires reported to U.S. municipal fire departments and so exclude fires reported only to federal or state agencies or industrial fire brigades. The 2007–2011 annual averages in this analysis are projections based on the detailed coded information collected in Version 5.0 of the U.S. Fire Administration's National Fire Incident Reporting System (NFIRS 5.0) and the National Fire Protection Association's (NFPA's) annual fire department experience survey. Except for property use and incident type, fires with unknown or unreported data were allocated proportionally in calculations of national estimates. Casualty and loss projections can be heavily influenced by the inclusion or exclusion of one or more unusually serious fires. Property damage has not been adjusted for inflation. Fires, civilian deaths and injuries are rounded to the nearest one and direct property damage is rounded to the nearest hundred thousand dollars (US).

The following property type use codes were included:

- Public assembly (100–199)
- Educational (200–299)
- Health care, nursing homes, detention and correction (300–399)
- Residential, excluding unclassified (other residential) and one-or two-family homes (420–499). This includes hotels and motels, dormitories, residential board and care or assisted living, and rooming or boarding houses.
- Mercantile (500–589)
- Office buildings, including banks, veterinary or research offices, and post offices (590–599)
- Laboratories and data centres (629, 635, and 639)

© Fire Protection Research Foundation 2015
N. White, M. Delichatsios, *Fire Hazards of Exterior Wall Assemblies Containing Combustible Components*, SpringerBriefs in Fire, DOI 10.1007/978-1-4939-2898-9

- Manufacturing or processing (700)
- Selected storage properties: Refrigerated warehouses, warehouses, other vehicle storage, general vehicle parking garages, and fire stations (839, 880, 882, 888, and 891)

Note—Building fires are identified by NFIRS incident type 111.
Separate queries were performed for:

- Fires starting in or the exterior wall surface area (area of origin code 76),
- Fires that did not start on the exterior wall area but the item first ignited was an exterior sidewall covering, surface or finish, including eaves (item first ignited code 12); and for
- Fires which did not start in the exterior wall or area or with the ignition of exterior sidewall covering but fire spread beyond the object of origin (fire spread codes 2–5) and the item contributing most to fire spread was the exterior sidewall covering (item contributing to flame spread code 12).

Results were summed after unknown or missing data, including extent of fire spread for the last condition, were allocated. This summed result is taken to represent the total number of exterior wall fires.
Separate queries were performed for four above ground height groupings:

- 1–2 stories,
- 3–5 stories,
- 6–10 stories, and
- 11–100 stories.

Separate queries were performed for four categories of automatic extinguishing system (AES) presence:

- Present [code 1],
- Partial system present [code 2],
- NFPA adjustment indicating AES presence but the reason for failure was the AES was not in the fire area [converted to code 8], and
- None present [code N]

# Total Structure Fires in Selected Properties, by Property Use

**Table A1**  Total structure fires (NFIRS incident type 110–129) in selected occupancies, including fires with confined fire incident types, non-building structure fires, and fires in portable buildings

| Property use | Fires | Civilian deaths | Civilian injuries | Property damage (US$ millions) | Portion of total fires |
|---|---|---|---|---|---|
| Public assembly | 15,374 | 6 | 172 | $446.2 | (9%) |
| Educational | 6012 | 0 | 90 | $105.1 | (3%) |
| Institutional | 7153 | 6 | 182 | $59.6 | (4%) |
| Residential | 121,651 | 485 | 4592 | $1548.8 | (68%) |
| Mercantile | 15,198 | 20 | 287 | $724.8 | (9%) |
| Office building | 3538 | 4 | 40 | $112.1 | (2%) |
| Laboratory and Data centre | 234 | 0 | 10 | $22.5 | (0%) |
| Manufacturing or processing | 5742 | 8 | 176 | $593.2 | (3%) |
| Selected storage occupancies | 2930 | 8 | 45 | $230.7 | (2%) |
| Total | 177,833 | 537 | 5595 | $3842.9 | (100%) |

# Building Fires in Selected Properties that Began on, at or with an Exterior Wall, by Property Use

**Table A2**  Building fires (NFIRS incident type 111) in selected properties in which the area of origin, item first ignited or item contributing most to flame spread was an exterior wall, by property use

| Property use | Fires | Civilian deaths | Civilian injuries | Property damage (US$ millions) | Portion of total structure fires |
|---|---|---|---|---|---|
| Public assembly | 706 | 0 | 6 | $30.8 | (5%) |
| Educational | 127 | 0 | 0 | $2.8 | (2%) |
| Institutional | 94 | 0 | 0 | $4.6 | (1%) |
| Residential | 2889 | 18 | 133 | $197.2 | (2%) |
| Mercantile | 891 | 0 | 5 | $31.1 | (6%) |
| Office building | 210 | 0 | 3 | $7.6 | (6%) |
| Laboratory and Data centre | 5 | 0 | 0 | $1.5 | (2%) |
| Manufacturing or processing | 120 | 0 | 1 | $6.3 | (2%) |
| Selected storage occupancies | 303 | 0 | 0 | $13.1 | (10%) |
| Total | 5346 | 18 | 148 | $295.0 | (3%) |

## Building Fires in Which the Area of Origin, Item First Ignited or Item Contributing Most to Flame Spread Was an Exterior Wall with the Different Height Groupings

**Table A3** Building fires (NFIRS incident type 111) in selected properties 1–2 stories in height in which the area of origin, item first ignited or item contributing most to flame spread was an exterior wall, by property use

| Property use | Fires | Civilian deaths | Civilian injuries | Property damage (US$ millions) |
|---|---|---|---|---|
| Public assembly | 660 | 0 | 5 | $26.7 |
| Educational | 115 | 0 | 0 | $2.7 |
| Institutional | 79 | 0 | 0 | $1.6 |
| Residential | 1948 | 17 | 72 | $111.2 |
| Mercantile | 828 | 0 | 4 | $29.9 |
| Office building | 187 | 0 | 0 | $7.2 |
| Laboratory and Data centre | 5 | 0 | 0 | $1.5 |
| Manufacturing or processing | 109 | 0 | 1 | $5.7 |
| Selected storage occupancies | 295 | 0 | 0 | $13.0 |
| Total | 4225 | 17 | 82 | $199.4 |

**Table A4** Building fires (NFIRS incident type 111) in selected properties 3–5 stories in height in which the area of origin, item first ignited or item contributing most to flame spread was an exterior wall, by property use

| Property use | Fires | Civilian deaths | Civilian injuries | Property damage (US$ millions) |
|---|---|---|---|---|
| Public assembly | 36 | 0 | 0 | $3.7 |
| Educational | 10 | 0 | 0 | $0.1 |
| Institutional | 11 | 0 | 0 | $3.0 |
| Residential | 878 | 0 | 60 | $82.1 |
| Mercantile | 45 | 0 | 1 | $1.1 |
| Office building | 16 | 0 | 3 | $0.4 |
| Laboratory and Data centre | 0 | 0 | 0 | $0.0 |
| Manufacturing or processing | 9 | 0 | 0 | $0.6 |
| Selected storage occupancies | 4 | 0 | 0 | $0.0 |
| Total | 1009 | 1 | 64 | $91.0 |

**Table A5** Building fires (NFIRS incident type 111) in selected properties 6–10 stories in height in which the area of origin, item first ignited or item contributing most to flame spread was an exterior wall, by property use

| Property use | Fires | Civilian deaths | Civilian injuries | Property damage (US$ millions) |
|---|---|---|---|---|
| Public assembly | 4 | 0 | 0 | $0.0 |
| Educational | 1 | 0 | 0 | $0.0 |
| Institutional | 2 | 0 | 0 | $0.0 |
| Residential | 24 | 0 | 0 | $0.4 |
| Mercantile | 4 | 0 | 0 | $0.0 |
| Office building | 2 | 0 | 0 | $0.0 |
| Laboratory and Data centre | 0 | 0 | 0 | $0.0 |
| Manufacturing or processing | 0 | 0 | 0 | $0.0 |
| Selected storage occupancies | 3 | 0 | 0 | $0.0 |
| Total | 39 | 0 | 0 | $0.5 |

**Table A6** Building fires (NFIRS incident type 111) in selected properties 11–100 stories in height in which the area of origin, item first ignited or item contributing most to flame spread was an exterior wall, by property use

| Property use | Fires | Civilian deaths | Civilian injuries | Property damage (US$ millions) |
|---|---|---|---|---|
| Public assembly | 8 | 0 | 1 | $0.4 |
| Educational | 1 | 0 | 0 | $0.0 |
| Institutional | 2 | 0 | 0 | $0.0 |
| Residential | 39 | 0 | 1 | $3.5 |
| Mercantile | 14 | 0 | 0 | $0.2 |
| Office building | 7 | 0 | 0 | $0.0 |
| Laboratory and Data centre | 0 | 0 | 0 | $0.0 |
| Manufacturing or processing | 2 | 0 | 0 | $0.0 |
| Selected storage occupancies | 2 | 0 | 0 | $0.0 |
| Total | 73 | 0 | 1 | $4.2 |

## Building Fires in Which the Area of Origin, Item First Ignited or Item Contributing Most to Flame Spread Was an Exterior Wall with the Different Height Groupings and Aes Presence

**Table A7** Building fires (NFIRS incident type 111) in properties of one or two stories in which the area of origin, item first ignited or item contributing most to flame spread was an exterior wall and some type of automatic extinguishing system was present (Code 1), by property use

| Property use | Fires | Civilian deaths | Civilian injuries | Property damage (US$ millions) |
|---|---|---|---|---|
| Public assembly | 132 | 0 | 1 | $3.1 |
| Educational | 17 | 0 | 0 | $0.2 |
| Institutional | 26 | 0 | 0 | $0.4 |
| Residential | 80 | 0 | 3 | $2.3 |
| Mercantile | 340 | 0 | 2 | $5.3 |
| Office building | 141 | 0 | 0 | $5.9 |
| Laboratory and Data centre | 0 | 0 | 0 | $0.0 |
| Manufacturing or processing | 28 | 0 | 0 | $0.6 |
| Selected storage occupancies | 194 | 0 | 0 | $7.8 |
| Total | 958 | 0 | 6 | $25.5 |

**Table A8** Building fires (NFIRS incident type 111) in properties of one or two stories in which the area of origin, item first ignited or item contributing most to flame spread was an exterior wall and a partial automatic extinguishing system was present (Code 2), by property use

| Property use | Fires | Civilian deaths | Civilian injuries | Property damage (US$ millions) |
|---|---|---|---|---|
| Public assembly | 15 | 0 | 0 | $1.9 |
| Educational | 0 | 0 | 0 | $0.0 |
| Institutional | 0 | 0 | 0 | $0.0 |
| Residential | 4 | 0 | 0 | $1.1 |
| Mercantile | 12 | 0 | 0 | $1.4 |
| Office building | 0 | 0 | 0 | $0.0 |
| Laboratory and Data centre | 0 | 0 | 0 | $0.0 |
| Manufacturing or processing | 6 | 0 | 0 | $0.0 |
| Selected storage occupancies | 5 | 0 | 0 | $0.0 |
| Total | 42 | 0 | 0 | $4.3 |

**Table A9** Building fires (NFIRS incident type 111) in properties of one or two stories in which the area of origin, item first ignited or item contributing most to flame spread was an exterior wall and an automatic extinguishing system was present outside of the fire area and failed to operate (Code 8), by property use

| Property use | Fires | Civilian deaths | Civilian injuries | Property damage (US$ millions) |
|---|---|---|---|---|
| Public assembly | 17 | 0 | 0 | $0.7 |
| Educational | 1 | 0 | 0 | $0.0 |
| Institutional | 1 | 0 | 0 | $0.1 |
| Residential | 11 | 0 | 1 | $1.9 |
| Mercantile | 47 | 0 | 0 | $0.8 |
| Office building | 11 | 0 | 0 | $0.0 |
| Laboratory and Data centre | 1 | 0 | 0 | $0.8 |
| Manufacturing or processing | 1 | 0 | 0 | $0.0 |
| Selected storage occupancies | 30 | 0 | 0 | $4.3 |
| Total | 120 | 0 | 1 | $8.6 |

**Table A10** Building fires (NFIRS incident type 111) in properties of one or two stories in which the area of origin, item first ignited or item contributing most to flame spread was an exterior wall and no automatic extinguishing system was present (Code N), by property use

| Property use | Fires | Civilian deaths | Civilian injuries | Property damage (US$ millions) |
|---|---|---|---|---|
| Public assembly | 495 | 0 | 4 | $21.2 |
| Educational | 97 | 0 | 0 | $2.5 |
| Institutional | 52 | 0 | 0 | $1.1 |
| Residential | 1854 | 17 | 68 | $105.8 |
| Mercantile | 429 | 0 | 2 | $22.5 |
| Office building | 35 | 0 | 0 | $1.3 |
| Laboratory and Data centre | 4 | 0 | 0 | $0.7 |
| Manufacturing or processing | 74 | 0 | 1 | $5.1 |
| Selected storage occupancies | 66 | 0 | 0 | $0.9 |
| Total | 3106 | 17 | 75 | $161.0 |

**Table A11** Building fires (NFIRS incident type 111) in properties of 3–5 stories in which the area of origin, item first ignited or item contributing most to flame spread was an exterior wall and some type of automatic extinguishing system was present (Code 1), by property use

| Property use | Fires | Civilian deaths | Civilian injuries | Property damage (US$ millions) |
|---|---|---|---|---|
| Public assembly | 13 | 0 | 0 | $2.5 |
| Educational | 3 | 0 | 0 | $0.0 |
| Institutional | 4 | 0 | 0 | $0.0 |
| Residential | 119 | 0 | 3 | $20.4 |
| Mercantile | 5 | 0 | 0 | $0.0 |
| Office building | 1 | 0 | 0 | $0.0 |
| Laboratory and Data centre | 0 | 0 | 0 | $0.0 |
| Manufacturing or processing | 3 | 0 | 0 | $0.5 |
| Selected storage occupancies | 0 | 0 | 0 | $0.0 |
| Total | 149 | 0 | 3 | $23.4 |

**Table A12** Building fires (NFIRS incident type 111) in properties of 3–5 stories in which the area of origin, item first ignited or item contributing most to flame spread was an exterior wall and a partial automatic extinguishing system was present (Code 2), by property use

| Property use | Fires | Civilian deaths | Civilian injuries | Property damage (US$ millions) |
|---|---|---|---|---|
| Public assembly | 1 | 0 | 0 | $0.0 |
| Educational | 0 | 0 | 0 | $0.0 |
| Institutional | 0 | 0 | 0 | $0.0 |
| Residential | 12 | 0 | 1 | $5.0 |
| Mercantile | 0 | 0 | 0 | $0.0 |
| Office building | 0 | 0 | 0 | $0.0 |
| Laboratory and Data centre | 0 | 0 | 0 | $0.0 |
| Manufacturing or processing | 0 | 0 | 0 | $0.0 |
| Selected storage occupancies | 0 | 0 | 0 | $0.0 |
| Total | 12 | 0 | 1 | $5.0 |

**Table A13** Building fires (NFIRS incident type 111) in properties of 3–5 stories in which the area of origin, item first ignited or item contributing most to flame spread was an exterior wall and an automatic extinguishing system was present outside of the fire area and failed to operate (Code 8), by property use

| Property use | Fires | Civilian deaths | Civilian injuries | Property damage (US$ millions) |
|---|---|---|---|---|
| Public assembly | 0 | 0 | 0 | $0.0 |
| Educational | 1 | 0 | 0 | $0.0 |
| Institutional | 2 | 0 | 0 | $0.0 |
| Residential | 22 | 0 | 3 | $5.9 |
| Mercantile | 1 | 0 | 0 | $0.0 |
| Office building | 0 | 0 | 0 | $0.0 |
| Laboratory and Data centre | 0 | 0 | 0 | $0.0 |
| Manufacturing or processing | 0 | 0 | 0 | $0.0 |
| Selected storage occupancies | 0 | 0 | 0 | $0.0 |
| Total | 26 | 0 | 3 | $5.9 |

**Table A14** Building fires (NFIRS incident type 111) in properties of 3–5 stories in which the area of origin, item first ignited or item contributing most to flame spread was an exterior wall and no automatic extinguishing system was present (Code N), by property use

| Property use | Fires | Civilian deaths | Civilian injuries | Property damage (US$ millions) |
|---|---|---|---|---|
| Public assembly | 22 | 0 | 0 | $1.2 |
| Educational | 7 | 0 | 0 | $0.1 |
| Institutional | 5 | 0 | 0 | $2.9 |
| Residential | 725 | 0 | 53 | $50.8 |
| Mercantile | 39 | 0 | 1 | $1.0 |
| Office building | 14 | 0 | 3 | $0.4 |
| Laboratory and Data centre | 0 | 0 | 0 | $0.0 |
| Manufacturing or processing | 5 | 0 | 0 | $0.1 |
| Selected storage occupancies | 4 | 0 | 0 | $0.0 |
| Total | 821 | 1 | 57 | $56.6 |

**Table A15** Building fires (NFIRS incident type 111) in properties of 6–10 stories in which the area of origin, item first ignited or item contributing most to flame spread was an exterior wall and some type of automatic extinguishing system was present (Code 1), by property use

| Property use | Fires | Civilian deaths | Civilian injuries | Property damage (US$ millions) |
|---|---|---|---|---|
| Public assembly | 2 | 0 | 0 | $0.0 |
| Educational | 0 | 0 | 0 | $0.0 |
| Institutional | 1 | 0 | 0 | $0.0 |
| Residential | 9 | 0 | 0 | $0.1 |
| Mercantile | 1 | 0 | 0 | $0.0 |
| Office building | 2 | 0 | 0 | $0.0 |
| Laboratory and Data centre | 0 | 0 | 0 | $0.0 |
| Manufacturing or processing | 0 | 0 | 0 | $0.0 |
| Selected storage occupancies | 0 | 0 | 0 | $0.0 |
| Total | 15 | 0 | 0 | $0.1 |

**Table A16** Building fires (NFIRS incident type 111) in properties of 6–10 stories in which the area of origin, item first ignited or item contributing most to flame spread was an exterior wall and a partial automatic extinguishing system was present (Code 2), by property use

| Property use | Fires | Civilian deaths | Civilian injuries | Property damage (US$ millions) |
|---|---|---|---|---|
| Public assembly | 0 | 0 | 0 | $0.0 |
| Educational | 0 | 0 | 0 | $0.0 |
| Institutional | 0 | 0 | 0 | $0.0 |
| Residential | 0 | 0 | 0 | $0.0 |
| Mercantile | 0 | 0 | 0 | $0.0 |
| Office building | 0 | 0 | 0 | $0.0 |
| Laboratory and Data centre | 0 | 0 | 0 | $0.0 |
| Manufacturing or processing | 0 | 0 | 0 | $0.0 |
| Selected storage occupancies | 0 | 0 | 0 | $0.0 |
| Total | 0 | 0 | 0 | $0.0 |

**Table A17** Building fires (NFIRS incident type 111) in properties of 6–10 stories in which the area of origin, item first ignited or item contributing most to flame spread was an exterior wall and an automatic extinguishing system was present outside of the fire area and failed to operate (Code 8), by property use

| Property use | Fires | Civilian deaths | Civilian injuries | Property damage (US$ millions) |
|---|---|---|---|---|
| Public assembly | 0 | 0 | 0 | $0.0 |
| Educational | 0 | 0 | 0 | $0.0 |
| Institutional | 0 | 0 | 0 | $0.0 |
| Residential | 0 | 0 | 0 | $0.2 |
| Mercantile | 0 | 0 | 0 | $0.0 |
| Office building | 0 | 0 | 0 | $0.0 |
| Laboratory and Data centre | 0 | 0 | 0 | $0.0 |
| Manufacturing or processing | 0 | 0 | 0 | $0.0 |
| Selected storage occupancies | 0 | 0 | 0 | $0.0 |
| Total | 1 | 0 | 0 | $0.2 |

**Table A18** Building fires (NFIRS incident type 111) in properties of 6–10 stories in which the area of origin, item first ignited or item contributing most to flame spread was an exterior wall and no automatic extinguishing system was present (Code N), by property use

| Property use | Fires | Civilian deaths | Civilian injuries | Property damage (US$ millions) |
|---|---|---|---|---|
| Public assembly | 1 | 0 | 0 | $0.0 |
| Educational | 1 | 0 | 0 | $0.0 |
| Institutional | 1 | 0 | 0 | $0.0 |
| Residential | 14 | 0 | 0 | $0.0 |
| Mercantile | 2 | 0 | 0 | $0.0 |
| Office building | 0 | 0 | 0 | $0.0 |
| Laboratory and Data centre | 0 | 0 | 0 | $0.0 |
| Manufacturing or processing | 0 | 0 | 0 | $0.0 |
| Selected storage occupancies | 2 | 0 | 0 | $0.0 |
| Total | 22 | 0 | 0 | $0.1 |

**Table A19** Building fires (NFIRS incident type 111) in properties of 11–100 stories in which the area of origin, item first ignited or item contributing most to flame spread was an exterior wall and some type of automatic extinguishing system was present (Code 1), by property use

| Property use | Fires | Civilian deaths | Civilian injuries | Property damage (US$ millions) |
|---|---|---|---|---|
| Public assembly | 1 | 0 | 0 | $0.0 |
| Educational | 0 | 0 | 0 | $0.0 |
| Institutional | 1 | 0 | 0 | $0.0 |
| Residential | 4 | 0 | 0 | $0.0 |
| Mercantile | 6 | 0 | 0 | $0.1 |
| Office building | 2 | 0 | 0 | $0.0 |
| Laboratory and Data centre | 0 | 0 | 0 | $0.0 |
| Manufacturing or processing | 1 | 0 | 0 | $0.0 |
| Selected storage occupancies | 0 | 0 | 0 | $0.0 |
| Total | 14 | 0 | 0 | $0.1 |

**Table A20** Building fires (NFIRS incident type 111) in properties of 11–100 stories in which the area of origin, item first ignited or item contributing most to flame spread was an exterior wall and a partial automatic extinguishing system was present (Code 2), by property use

| Property use | Fires | Civilian deaths | Civilian injuries | Property damage (US$ millions) |
|---|---|---|---|---|
| Public assembly | 1 | 0 | 0 | $0.0 |
| Educational | 0 | 0 | 0 | $0.0 |
| Institutional | 0 | 0 | 0 | $0.0 |
| Residential | 1 | 0 | 0 | $2.1 |
| Mercantile | 0 | 0 | 0 | $0.0 |
| Office building | 1 | 0 | 0 | $0.0 |
| Laboratory and Data centre | 0 | 0 | 0 | $0.0 |
| Manufacturing or processing | 0 | 0 | 0 | $0.0 |
| Selected storage occupancies | 0 | 0 | 0 | $0.0 |
| Total | 2 | 0 | 0 | $2.1 |

**Table A21** Building fires (NFIRS incident type 111) in properties of 11–100 stories in which the area of origin, item first ignited or item contributing most to flame spread was an exterior wall and an automatic extinguishing system was present outside of the fire area and failed to operate (Code 8), by property use

| Property use | Fires | Civilian deaths | Civilian injuries | Property damage (US$ millions) |
|---|---|---|---|---|
| Public assembly | 1 | 0 | 0 | $0.0 |
| Educational | 0 | 0 | 0 | $0.0 |
| Institutional | 0 | 0 | 0 | $0.0 |
| Residential | 2 | 0 | 0 | $0.0 |
| Mercantile | 0 | 0 | 0 | $0.0 |
| Office building | 0 | 0 | 0 | $0.0 |
| Laboratory and Data centre | 0 | 0 | 0 | $0.0 |
| Manufacturing or processing | 0 | 0 | 0 | $0.0 |
| Selected storage occupancies | 0 | 0 | 0 | $0.0 |
| Total | 3 | 0 | 0 | $0.0 |

**Table A22** Building fires (NFIRS incident type 111) in properties of 11–100 stories in which the area of origin, item first ignited or item contributing most to flame spread was an exterior wall and no automatic extinguishing system was present (Code N), by property use

| Property use | Fires | Civilian deaths | Civilian injuries | Property damage (US$ millions) |
|---|---|---|---|---|
| Public assembly | 5 | 0 | 1 | $0.4 |
| Educational | 1 | 0 | 0 | $0.0 |
| Institutional | 0 | 0 | 0 | $0.0 |
| Residential | 32 | 0 | 1 | $1.3 |
| Mercantile | 8 | 0 | 0 | $0.2 |
| Office building | 4 | 0 | 0 | $0.0 |
| Laboratory and Data centre | 0 | 0 | 0 | $0.0 |
| Manufacturing or processing | 1 | 0 | 0 | $0.0 |
| Selected storage occupancies | 2 | 0 | 0 | $0.0 |
| Total | 53 | 0 | 1 | $2.0 |

# Appendix B: Regulations—Detailed Summaries

*Australian National Construction Code*

The Australian National Construction Code (NCC) classifies buildings as follows

- Class 1—single dwellings. For example residential houses
- Class 2—building containing 2 or more sole occupancy units. For example apartment building
- Class 3—Residential building other than class 1 or 2 which provides long term or transient living for a number of unrelated occupants. For example boarding house, hotel, motel.
- Class 4—a dwelling in a building that is class 5, 6, 7, 8 or 9 that is the only dwelling within the building.
- Class 5—office building
- Class 6—shop or retail building
- Class 7—a building which is

  - Class 7a—car park
  - Class 7b—storage or warehouse

- Class 8—laboratory or workshop or factory
- Class 9—public building which is

  - Class 9a—health care building/hospital
  - Class 9b—Assembly building. For example educational, public halls, cinemas, nightclubs
  - Class 9c—Aged care

- Class 10—non-habitable structure. For example shed

The National Construction Code is a performance based code which specifies prescriptive requirements called Deemed-to-Satisfy (DtS) requirements and also

© Fire Protection Research Foundation 2015

N. White, M. Delichatsios, *Fire Hazards of Exterior Wall Assemblies Containing Combustible Components*, SpringerBriefs in Fire, DOI 10.1007/978-1-4939-2898-9

permits performance based alternative solutions provided that these alternative solutions are demonstrated by fire engineering analysis to satisfy the codes performance requirements.

National Construction Code Volume 1 pertains to class 2–9 buildings. Volume 2 pertains to class 1 and 10 buildings which are beyond the scope of this book.

CP2 and CP4 are the relevant performance requirements.


**NCC Prescriptive Requirements for Exterior Wall Materials**

The minimum type of fire resisting construction required is grouped into 3 different Types dependant on building class and rise in storeys as summarised in Table B1.

**Table B1**  Type of fire resisting construction

| Rise in storeys | Class of building | |
|---|---|---|
|  | 2, 3, 9 | 5, 6, 7, 8 |
| 4 or more | A | A |
| 3 | A | B |
| 2 | B | C |
| 1 | C | C |

Type A is highest level of fire resistant construction and Type C is the lowest level of fire resisting construction. Specific fire resistance levels are specified for different building elements, for each building class within each Type of construction.

NCC Vol 1 Specification C1.1 states that Type A and Type B construction requires external walls to be non combustible construction. Non-combustible is defined by either a non combustibility test of specific materials such as plasterboard and cement sheet which are deemed to be non-combustible. Effectively all residential and public buildings of buildings of two stories or greater and all other classes of building of three stories or greater are not permitted to have combustible facades. A concession does permit class 2 buildings of 2–3 stories to have external walls of light weight timber framed construction provided all other components of the external wall system are non-combustible and an automatic smoke alarm system is fitted to the building.

No restrictions to flammability of exterior wall systems are prescribed for buildings below these height limits. Australia does not specify a small or large scale fire test to determine the suitable fire performance of external wall systems other than the non-combustibility test.


**Fire Stop Barriers**

NCC Vol 1 Clause C2.6 states that buildings of Type A construction require any gaps behind curtain or panel walls at each floor level to be packed with a non-combustible material which is resistant to thermal or structural movement to act as a seal against fire or smoke. Fire stop barriers to external insulation systems are not prescribed as such materials are not permitted as DtS as external insulation systems are not permitted as DtS to buildings of three stories or greater.

## Separation Between Buildings

NCC Vol 1 Clause C3.2 states the requirements for protection of openings in external walls which requires that buildings are generally required to be separated from other buildings or fire source features by the following horizontal distances.

- 3 m from a side or rear boundary of an allotment
- 6 m from the far boundary of road, river, lake or the like adjoining the allotment
- 6 m from another building on the same allotment

If buildings are not separated by the above distances then buildings must be separated by walls having prescribed FRLs and all openings are to be protected by either external sprinkler protection or self closing barriers having prescribed FRL's.

## Separation of Vertical Openings

NCC Vol 1 Clause C2.6 states for Buildings of Type A construction, openings (windows) in external walls that are above openings in the storey below must be separated by either:

- A spandrel having an FRL of 60/60/60 that is at least 900 mm in height and extends at least 600 mm above the intervening floor, or
- A horizontal projection having an FRL of 60/60/60 which projects 1100 mm horizontally from the external face of the wall and extends along the wall at least 450 mm beyond the openings.

The above separation is not required if the building is internally sprinkler protected.

## Sprinkler Protection

NCC Vol 1 Clause E1.5 states that sprinkler protection is required throughout an entire building for buildings with an effective height greater than 25 m or for buildings where maximum fire compartment size limits (which are dependent on the class of building) are exceeded.

# New Zealand Building Code

The New Zealand Building Code is a performance based building code which specifies prescriptive requirements called Acceptable Solutions (AS) but also permits performance based alternative solutions provided that these alternative solutions are demonstrated by fire engineering analysis to satisfy the codes performance requirements.

Acceptable solution (prescriptive requirements) are detailed in the separate documents as listed in the following table for different types of buildings.

**Table B2**  New Zealand Acceptable solution documents for different building types

| Acceptable solution document | Building type | Applies to | Comment |
|---|---|---|---|
| C/AS1 | Single household units and small multi-unit dwellings | Houses, townhouses and small *multi-unit dwellings* | Outside scope of this book |
|  |  | Limited area outbuildings |  |
| C/AS2 | Sleeping (non institutional) | Permanent accommodation e.g. apartments |  |
|  |  | Transient accommodation e.g. hotels, motels, hostels, backpackers |  |
|  |  | Education accommodation |  |
| C/AS3 | Care or detention | Institutions, hospitals (excluding special care facilities), residential care, resthomes, medical day treatment (using sedation), detention facilities (excluding prisons) |  |
| C/AS4 | Public access and educational facilities | Crowds, halls, recreation centres, public libraries (<2.4 m storage height), cinemas, shops, personal services (e.g. dentists and doctors except as included above, beautician and hairdressing salons), schools, restaurants and cafes, *early childhood centres* |  |
| C/AS5 | Business, commercial and low level storage | Offices (including professional services such as law and accountancy practices), laboratories, workshops, manufacturing (excluding *foamed plastics*), factories, processing, cool stores (capable of <3.0 m storage height) and warehouses and other storage units capable of <5.0 m storage height, light aircraft hangars |  |
| C/AS6 | High level storage and other high risks | Warehouses (capable of <5.0 m storage height), cool stores (capable of <3.0 m storage height), trading and bulk retail (<3.0 m storage height) |  |
| C/AS7 | Vehicle storage and parking | Vehicle parking—within a *building* or a separate *building* | Outside scope of this book |

## Requirements for Exterior Wall Materials

The acceptable level of fire performance of external wall systems depends on the building height, presence of sprinklers and the distance from the relevant boundary of the allotment.

**Table B3**  NZ Building code requirements for exterior wall fire performance

| Building type | Requirements | |
|---|---|---|
| | Distance to boundary and building height | Cone Calorimeter test requirements at irradiance of 50 kW/m$^2$ for duration of 15 min |
| Sleeping/Residential (non institutional) AS2 | Distance to relevant boundary <1.0 m | Peak HRR shall not exceed 100 kW/m$^2$ and total heat released shall not exceed 25 MJ/m$^2$ |
| Public access and educational facilities AS4 | | |
| Business, commercial and low level storage AS5 | | |
| High level storage and other high risks AS6 | Distance to relevant boundary ≥1.0 m and building height >7.0 m | Peak HRR shall not exceed 150 kW/m$^2$ and total heat released shall not exceed 50 MJ/m$^2$ |
| Care or detention (hospitals or prisons AS3 | Distance to relevant boundary <1.0 m, or building height >7.0 m | Peak HRR shall not exceed 100 kW/m$^2$ and total heat released shall not exceed 25 MJ/m$^2$ |
| | Distance to relevant boundary ≥1.0 m and building height ≤7.0 m | Peak HRR shall not exceed 150 kW/m$^2$ and total heat released shall not exceed 50 MJ/m$^2$ |

However the requirements in Table B3 do not apply if:

(a) *Surface finishes* are no more than 1 mm in thickness and applied directly to a *non-combustible* substrate, or
(b) The entire wall assembly has been tested at full scale in accordance with NFPA 285 and has passed the test criteria, or
(c) The *building* is sprinklered and has a *building height* of 25 m or less

## Fire Stop Barriers

Fire stopping is required for all interior gaps at fire compartment (fire cell) boundaries. This includes gaps between slabs and exterior wall systems such as curtain walls. The fire stopping must have a fire resistance rating equivalent to that required for the fire compartment boundary.

Mineral wool fire stop barriers (at least 50 mm thick) are required for buildings of three or more stories fitted with combustible external insulation. The fire stop barriers must be installed to the cladding at intervals of not more than two stories.

**Separation Between Buildings**

The critical distance for separation of buildings from the boundary in terms of protection of openings and fire performance of external cladding is 1 m. At less than 1 m separation all openings (windows) must be protected by fire rated glass. At greater than 1 m the percentage of unprotected opening area permitted for external walls gradually increases with no requiring for protection at a separation distances ranging from 6 m for residential buildings (AS2) to 16 m for high risk storage (AS6).

**Separation of Vertical Openings**

Openings (windows) in external walls that are above openings in the fire compartment below must be separated by a combination of spandrels and/or horizontal projections having the same FRL as the floor separating the upper and lower fire compartments.

**Table B4** Permitted combinations of horizontal projection and spandrel separation of openings

| Horizontal projection (m) | Spandrel height (m) |
|---|---|
| 0.0 | 1.5 |
| 0.3 | 1.0 |
| 0.45 | 0.5 |
| 0.6 | 0.0 |

The above separation of vertical openings is not required where the building is internally sprinkler protected.

**Sprinkler Protection**

Sprinkler protection is generally required for most building types where the height exceeds 25 m or where maximum compartment size limits are exceeded. Sprinkler protection is generally required for all care or detention type buildings.

# UK Approved Document B

The Building Regulations 2010 for England and Wales state the performance requirements with regards to fire safety. Approved Document B is a guidance documents which states prescriptive requirements for fire safety which achieve compliance with the Building Regulations 2010. Alternative solutions supported by fire engineering analysis are permitted.

## Reaction to Fire Requirements for Exterior Wall Materials

Approved Document B, Section 12 states external wall construction should either meet the limited combustibility requirements in Table B5 or should meet the performance requirements given in BRE report BR 135 using full scale test data from BS 8414-1 or BS 8414-2.

**Table B5** Approved Document B limited combustibility requirements for external wall construction

|     | Type of building | Building height (m) | Distance from relevant boundary (m) | Reaction to fire requirements |
| --- | --- | --- | --- | --- |
| (a) | Any building | <18 | <1 | All exterior walls >1 m from boundary to be either Class 0 (national class) or Class B-s3,d2 or better (Euroclass) |
| (b) | Any building except (c) | <18 | ≥1 | No requirements |
| (c) | Assembly or recreation building of more than one storey | <18 | ≥1 | All exterior walls up to 10 m above ground level or a roof or any other external part of the building accessible by the public to be either Class 0 (national class) or Class B-s3,d2 or better (Euroclass) |
| (d) | Any building | ≥18 | <1 | either Class 0 (national class) or Class B-s3,d2 or better (Euroclass) |
| (e) | Any building | ≥18 | ≥1 | External wall up to 18 m above ground level to be Index (I) not more than 20 (national class) or Class C-s3,d2 or better (Euroclass) |
|     |              |     |    | External walls 18 m and above to be either Class 0 (national class) or Class B-s3,d2 or better (Euroclass) |

UK National Class 0 materials are either non combustible when tested to BS 476-4 or Limited combustibility when tested to BS 476-11.

Index (I) is determined by testing to BS 476-6.

Euroclass refers to classification in accordance with EN 13501-1.

## Fire Stop Barriers

Cavity barriers having at least 30 min fire resistance must be provided to close the edges of cavities around openings (e.g. windows) and also within any wall cavities (internal or external) located at the junction of compartment floors or walls.

Internal gaps (e.g. between compartment floors the inside face of a wall such as a curtain wall) must be fire stopped with a material having a fire resistance at least equivalent to the compartment.

## Separation Between Buildings

The critical distance for separation of buildings from the boundary in terms of protection of openings and fire performance of external cladding is 1 m. At less than 1 m separation all openings (windows) must be protected by fire rated glass. At greater than 1 m the percentage of unprotected opening area permitted for external walls gradually increases to 100 % at a separation distances of 6 for small residential buildings, 12.5 m for larger residential, office, assembly and recreation and 25 for retail/commercial, industrial, storage and other non-residential type buildings.

## Separation of Vertical Openings

There is no requirement for vertical separation of openings in external walls between each level.

## Sprinkler Protection

Sprinkler protection is generally required for all building types excluding institutional, other residential and car parks where the height exceeds 30 m or where maximum compartment size limits are exceeded (as detailed in Table A2 of Approved Document B).

# Façade Regulations in Nordic Countries

Strömgren et al. [85] have provided a comparative analysis of façade regulations in Nordic countries. This analysis was based on a reference building of four stories which is considered to be a typical Nordic building. The following summaries of acceptable solution requirements are taken from Strömgren et al.

## Requirements for Exterior Wall Materials

The reaction to fire requirements for exterior wall materials in Nordic countries generally apply Euroclassifications as summarised in Table B6. Acceptable solutions vary from non-combustible materials (A2-s1,d0) to only fulfilling variations of Euroclass

B. In Sweden, full-scale testing to SP Fire 105 is also accepted as an alternative. Some countries allow some parts of the façade to be of a lower class, i.e. D-s2,d0.

**Table B6** Nordic requirements for exterior wall reaction to fire

| Country | Protection against fire spread along the façade | Reaction to fire requirements for components in the external wall | Protection against falling objects |
|---|---|---|---|
| Sweden | A2-s1,d0 | A2-s1,d0 | Compliance can be shown by testing with SP Fire 105 |
|  | Certain exceptions allow D-s2,d2, for instance if sprinklers are installed in the building or only limited areas of the facade | or | |
|  |  | Fire stops preventing fire spread required at each floor unless the whole external wall | |
|  | or | or | |
|  | Compliance can be shown by testing with SP Fire 105 | Compliance can be shown by testing with SP Fire 105 | |
| Denmark | Covering class K1 10 B-s1, d0 or K1 10 D-s2 d2 (depending on building height) | See "Protection against fire spread along the façade" | No requirements |
|  | Certain exceptions allow D-s2,d2 for lower buildings | | |
|  | Insulation materials with D-s1,d0 or lower poorer than material class D-s2,d2 (material level) must be protected with a covering class K1 10 B-s1, d0 or a construction class EI/REI30 or a construction class EI/REI30 and A2-s1,d0 (depending on building height) on each side | | |
| Norway | Cladding of class B-s3,d0. However D-s3,d0 in low rise (maximum four stories, depending on risk class and hazard class) and if the fire risk in the facade is limited and the risk of fire spread to other buildings is low | Insulation must be of class A2-s1,d0 | No Specific requirements. Compliance can be shown by testing with SP Fire 105 |
|  |  | External insulation systems for existing building: Testing according to SP Fire 105. However not pre-accepted in hazard class 3 (more than four stories) and risk class 6 (hospitals, hotels etc.) | |

(continued)

**Table B6** (continued)

| Country | Protection against fire spread along the façade | Reaction to fire requirements for components in the external wall | Protection against falling objects |
|---------|------------------------------------------------|------------------------------------------------------------------|-----------------------------------|
| Finland | 3–8 floors (apartment and office buildings): B-s2,d0 generally and D-s2,d2 if building sprinklered (excluding first floor) | In designing the constructions of external walls, the hazard of fire spreading within the construction and through the joints shall be considered | Applies only when D-s2,d2 class cladding (wood) is used in 3–8 floor buildings |
|         | Higher buildings: B-s1,d0 +Certain exceptions allow D-s2, d2 for minor areas | P1 class buildings (number of floors: 3—unlimited): Thermal insulation which is inferior to class B–s1, d0 shall be protected and positioned in such a manner that the spread of fire into the insulation, from one fire compartment to another and from one building to another building is prevented. In these cases rendering or a metal sheet is generally not a sufficient protection | |
|         |                                                | Protected combustible insulation can be allowed in certain cases. For example coverings fulfilling fire resistance EI 30 or large scale or some experimental/calculation evidence on protective performance/no contribution to fire spread. A2-s1,d0 or B-s2, d0 if the load bearing construction is combustible (buildings with 3–8 floors) | |

## Fire Stop Barriers

There is some variation between Nordic countries however Fire stop barriers are generally required at each floor between the slab and the rear/inside of the exterior wall. Where combustible exterior insulation is applied fire stops must generally be imbedded in the insulation at each floor level (unless suitable performance is demonstrated in the large scale SP105 test).

## Separation Between Buildings

Requirements relating to this item have not been determined.

## Separation of Vertical Openings

Separation distance between windows is only explicitly required in Sweden, which requires 1.2 m spandrel separation or windows with 30 min fire resistance. Norway has special requirements that is connected to fire resistance solutions. Finland has no requirements whereas Denmark requires a risk evaluation if the façade is sloping.

## Sprinkler Protection

Requirements relating to this item have not been determined.

# International Building Code (IBC), USA

The International Building Code (IBC) is a model building code developed by the International Code Council (ICC). It has been adopted throughout most of the United States. In many cases the IBC may only be adopted in part or with modifications in various States within America.

Buildings are classified into 5 different types of construction having a decreasing level of fire resistance in the following order; Type I, Type II, Type III, Type IV and Type V. Building classes having lower levels of fire resistance are limited to low building heights. Type V construction has the lowest fire resistance and is typically timber framed construction.

## Requirements for Exterior Wall Materials

The general performance requirement for combustible exterior wall systems is that for buildings of Type I, II, III or IV construction that are greater than 12.192 m in height must be tested and comply with NFPA 285 full scale façade test (IBC Section 1403.5).

However the IBC also gives the following detailed reaction to fire requirements for specific types of materials. It is presumed that if these specific requirements are met then demonstration of compliance with the NFPA 285 test is not required.

### Combustible Exterior Wall Coverings

Buildings of Type I, II, III or IV construction are permitted to have combustible exterior wall coverings if they meeting the following requirements

- Combustible coverings ≤10 % of exterior wall surface area where fire separation distance is ≤1.524 m
- Combustible coverings limited to 12.192 m in height

- Fire retardant treaded wood is not limited in area at any separation distance and is permitted up to 18.233 m in height
- Ignition resistance—combustible exterior wall coverings must be tested in accordance with NFPA 268 applying the following criteria (wood based products and combustible materials covered with a listed acceptable material of low combustibility are excluded)

  - Fire separation $\leq$1.524 m—combustible coverings shall not exhibit sustained flaming
  - Fire separation >1.524 m—the acceptable fire separation distance is dependent on the maximum radiant heat flux that does not cause sustained flaming and ranges from 1.524 m separation at 12.5 kW/m$^2$ to 7.62 at 3.5 kW/m$^2$.

Foam Plastic Insulation (ICC Section 2603)

Foam plastic insulation in or on exterior walls without a thermal barrier separation from the interior is permitted for one storey buildings with the following requirements:

- Flame spread index of $\leq$25 and a smoke developed index of $\leq$450 (ASTM E 84 or UL 723).
- Foam plastic thickness $\leq$102 mm
- Foam plastic covered by $\geq$0.81 mm aluminium or $\geq$0.41 mm steel
- Building must be sprinkler protected.

  Any Height

- Separated from building interior by approved thermal barrier 12.7 mm Gypsum wall board or equivalent.
- Insulation, exterior facings and coatings shall be tested separately to ASTM E 84 or UL 723 and shall have a flame spread index of $\leq$25 and a smoke developed index of $\leq$450. (aluminium composite panels of $\leq$6.4 mm are permitted to be tested as an assembly)
- Potential heat of foam plastic shall be determined applying NFPA 259. The potential heat of the foamed plastic in the installed walls shall not exceed that of the material tested in the full-scale façade test.
- The complete wall assembly must be tested and comply with NFPA 285 full-scale façade test

  Special Approval—Special approval may be provided without compliance with the above requirements based on large scale room corner tests such as NFPA 286, FM 4880, UL 1040 or UL 1715 if these tests are determined to be representative of the end use configuration.

## Light Transmitting Plastic Wall Panels (ICC Section 2607)

**Table B7** Summary ICC reaction to fire requirements for light transmitting plastic wall panels

| Type I, Type II, Type III and Type IV buildings | |
| --- | --- |
| Height | Requirement |
| Installed to a maximum height of 22.86 m (75 ft) or unlimited height if building is sprinkler protected | • Not permitted for building classes Assembly (A-1, A-2), High Hazard, Institutional (I-2, I-3) |
| | • Not permitted on exterior walls required to have a fire resistance rating (by other provisions of code) |
| | • Flame spread index of ≤75 and a smoke developed index of ≤450 (ASTM E 84 or UL 723) |
| | • Have a self ignition temperature ≥343 °C (tested to ASTM D 1929) |
| | • Be either CC1 (burn length ≤25 mm and self extinguishment) or CC2 (burning rate of ≤1.06 mm/min) when tested to ASTM D 635 |
| | • Than maximum area of exterior wall covered by plastic panels must be limited as stated in Table B10 or the maximum area of unprotected openings permitted (whichever is less). The maximum area of single plastic panels and minimum separation distance between panels must be limited as stated in Table B10 |
| | • For sprinkler protected buildings the maximum area of exterior wall covered and maximum area of single panels may be increased by 100 %. However maximum area of exterior wall covered must not exceed 50 % of the area of unprotected openings permitted (whichever is less) |
| Type V building | |
| Requirement for any height | Same as above except there is no limitation on area of coverage or required separation of panels |

## Fibre-Reinforced Polymer

**Table B8** Summary ICC reaction to fire requirements for fibre-reinforced polymer wall panel

| Height | Requirement |
| --- | --- |
| Installed to a maximum height of 12.19 m | • Comply with same requirements as for combustible exterior wall covering |
| | • Flame spread index of ≤200 (ASTM E 84 or UL 723) |
| | • Fire blocking of any concealed space in the exterior wall |
| Any height—Option 1 | • Comply with same requirements as for foam plastic insulation |
| | • Fire blocking of any concealed space in the exterior wall |
| Any height—Option 1 | • Cover <20 % of exterior wall area |
| | • Flame spread index of ≤25 (ASTM E 84 or UL 723) |
| | • Fire blocking of any concealed space in the exterior wall |
| | • Be installed directly to a non-combustible substrate or be separated from the exterior wall by steel (0.4 mm), aluminium (0.5 mm) or other approved non-combustible material |

## Metal Composite Materials (MCM) (Section 1407)

**Table B9**  Summary ICC reaction to fire requirements for MCM wall panels

| Type I, Type II, Type III and Type IV buildings | |
|---|---|
| Height | Requirement |
| Installed to a maximum height of 12.19 m (40 ft) | • Flame spread index of ≤75 and a smoke developed index of ≤450 (ASTM E 84 or UL 723) |
| | • Cover <10 % or exterior wall area where the horizontal separation from the boundary is ≤1525 mm, or |
| | • No Limit on area where. horizontal separation from the boundary is >1525 mm |
| Installed to a maximum height of 15.24 m (50 ft) | • Continuous areas of panels must not exceed 27.8 m² and must be separated from other continuous areas of panels by at least 1220 mm; and |
| | • Have a self ignition temperature ≥343 °C (tested to ASTM D 1929 standard test method for determining ignition temperature of plastics); and |
| | • Flame spread index of ≤75 and a smoke developed index of ≤450 (ASTM E 84 or UL 723) |
| Installed to a maximum height of 22.86 m (75 ft) or unlimited height if building is sprinkler protected | Option 1 |
| | • Not permitted for building classes A-1, A-2, H, I-2, I-3 |
| | • Not permitted on exterior walls required to have a fire resistance rating (by other provisions of code) |
| | • Flame spread index of ≤75 and a smoke developed index of ≤450 (ASTM E 84 or UL 723) |
| | • Have a self ignition temperature ≥343 °C (tested to ASTM D 1929) |
| | • Be either CC1 (burn length ≤25 mm and self extinguishment) or CC2 (burning rate of ≤1.06 mm/min) when tested to ASTM D 635 |
| | • Than maximum area of exterior wall covered by MCM panels must be limited as stated in Table B10 or the maximum area of unprotected openings permitted (whichever is less). The maximum area of single MCM panels and minimum separation distance between panels must be limited as stated in Table B10 |
| | • For sprinkler protected buildings the maximum area of exterior wall covered and maximum area of single panels may be increased by 100 %. However maximum area of exterior wall covered must not exceed 50 % of the area of unprotected openings permitted (whichever is less) |

| Type I, Type II, Type III and Type IV buildings | |
|---|---|
| Height | Requirement |
| | Option 2 |
| | • MCM must not be installed on any wall where separation distance <9.144 m or, Separation distance <6.096 m for sprinkler protected building |
| | • Flame spread index of ≤75 and a smoke developed index of ≤450 (ASTM E 84 or UL 723) |
| | • Have a self ignition temperature ≥343 °C (tested to ASTM D 1929) |
| | • Be either CC1 (burn length ≤25 mm and self extinguishment) or CC2 (burning rate of ≤1.06 mm/min) when tested to ASTM D 635 |
| | • The area of exterior wall covered shall be ≤25 %. The area of a single MCM panel one story or more above grade shall not exceed 1.5 m² and the vertical dimension of a single MCM panel shall not exceed 1.219 m |
| | • Vertical separation between panels shall be provided by flame barriers which extend 762 mm beyond the exterior wall or a vertical separation distance of 1.219 m |
| | • If the building is sprinkler protected then the area of exterior wall covered shall be ≤50 % and there is no limit to single panel size and no requirement for vertical separation of panels |
| Any height | • Compliance with NFPA 285 full scale façade test, And |
| | • Flame spread index of ≤25 and a smoke developed index of ≤450 (ASTM E 84 or UL 723) |
| | • Separated from building interior by approved thermal barrier 12.7 mm Gypsum wall board or equivalent. Thermal barrier not required if MCM system tested and approved to either UL 10 40 or UL 1715 |
| Type V building | |
| Requirement for any height | Flame spread index of ≤75 and a smoke developed index of ≤450 (ASTM E 84 or UL 723) |

**Table B10** ICC requirements for percentage of wall coverage, panel area and separation between panels for MCM or plastic panels

| Fire separation distance (feet) | Combustibility class of MCM or plastic wall panel | Maximum percentage area of Exterior Wall covered with MCM plastic panels | Maximum single area of MCM or plastic panels (square feet) | Minimum separation of MCM or plastic panels (feet) | |
|---|---|---|---|---|---|
| | | | | Vertical | Horizontal |
| <6 | – | Not permitted | Not permitted | – | – |
| 6 or more but <11 | CC1 | 10 | 50 | 8 | 4 |
| | CC2 | Not permitted | Not permitted | – | – |
| 11 or more but <30 | CC1 | 25 | 90 | 6 | 4 |
| | CC2 | 15 | 70 | 8 | 4 |
| >30 | CC1 | 50 | Not limited | 3 | 0 |
| | CC2 | 50 | 100 | 6 | 3 |

EIFS

EIFS must meet the requirements of ASTM E2568 [100]

High Pressure Laminates

High pressure laminates (HPL) must meet the following requirements (ICC Section 1409).

**Table B11** Summary ICC reaction to fire requirements for HPL wall panels

| Type I, Type II, Type III and Type IV buildings | |
| --- | --- |
| Height | Requirement |
| Installed to a maximum height of 12.19 m (40 ft) | Flame spread index of ≤75 and a smoke developed index of ≤450 (ASTM E 84 or UL 723) |
| | Cover <10 % or exterior wall area where the horizontal separation from the boundary is ≤1525 mm, or |
| | No Limit on area where. horizontal separation from the boundary is >1525 mm |
| Installed to a maximum height of 15.24 m (50 ft) | Continuous areas of panels must not exceed 27.8 m² and must be separated from other continuous areas of panels by at least 1220 mm; and |
| | Have a self ignition temperature ≥343 °C (tested to ASTM D 1929 standard test method for determining ignition temperature of plastics); and |
| | Flame spread index of ≤75 and a smoke developed index of ≤450 (ASTM E 84 or UL 723) |
| Any height | Compliance with NFPA 285 full scale façade test, And; |
| | Flame spread index of ≤25 and a smoke developed index of ≤450 (ASTM E 84 or UL 723) |
| | Separated from building interior by approved thermal barriers 12.7 mm Gypsum wall board or equivalent. Thermal barrier not required if HPL system tested and approved to either UL 10 40 or UL 1715 |
| Type V building | |
| Requirement for any height | Flame spread index of ≤75 and a smoke developed index of ≤450 (ASTM E 84 or UL 723) |

**Fire Stop Barriers**

Internal gaps (e.g. between compartment floors the inside face of a wall such as a curtain wall) must be fire stopped with an approved material having a fire resistance at least equivalent to the compartment (ICC Section 715).

Fire Blocking, using non combustible materials such as mineral wool is to be installed within concealed spaces of exterior wall coverings at maximum intervals of 6.096 m (both horizontally and vertically) so that the maximum concealed space does not exceed 9.3 m².

Use of fire stop barriers imbedded in EIFS may be specified in ASTM E2568.

## Separation Between Buildings

For non sprinkler protected buildings, no unprotected openings are permitted at a separation distance of less than 5 ft. The percentage of unprotected openings permitted increases to no limit at 30 ft.

For sprinkler protected buildings, no unprotected openings are permitted at a separation distance of less than 3 ft. The percentage of unprotected openings permitted increases to no limit at 20 ft.

## Separation of Vertical Openings

For buildings more than three stories in height which are not sprinkler protected openings must be separated from openings in the storey above by (IBC Section 705.8.5) either:

- the lower storey opening has a protection rating of at least 3/4 h, or
- A 915 mm spandrel with 1 h fire resistance, or
- A 760 mm horizontally projecting barrier with 1 h fire resistance.

## Sprinkler Protection

Typical thresholds above which sprinkler systems are required in the *International Building Code* (IBC) include:

- Mercantile: Over 12,000 ft² (1115 m²) in one fire area, or over 24,000 ft² (2230 m²) in combined fire area on all floors, or more than three stories in height
- High-Rise: All buildings over 75 ft (22.86 m) in height. However sprinklers are also required for all buildings with a floor level having an occupant load of 30 or more that is located over 55 ft (16.8 m) in height (IBC 903.2.11.3)
- Residential Apartments: All buildings except townhouses built as attached single-family dwellings

# NFPA 5000, USA

NFPA 5000 was developed as an alternative building code to the IBC. However in practice NFPA 5000 is not adopted by most states of America. The IBC is the model building code currently most adopted within the USA.

Buildings are classified into 5 different types of construction, the same as for the IBC.

NFPA 5000 Section 7.2 states that the general flammability requirement for all exterior walls for building class Type I, Type II, Type III and Type IV are required to meet the requirements of the large scale façade test NFPA 285.

However the following specific requirements for different types of exterior wall materials are also stated.

Foam plastic Insulation requirements are stated in NFPA 5000 Section 48.4.1. Foamed plastics used in exterior walls for Type I, Type II, Type III and Type IV buildings must comply all of the requirements in Table B12.

**Table B12** Foamed plastic insulation requirements for Type I, Type II, Type III and Type IV buildings

| Property | Requirement |
| --- | --- |
| Thermal barriers | Foam plastic insulation must be separated from the building by an acceptable thermal barrier such as 13 mm gypsum board or a material meeting temperature transmission and integrity requirements of NFPA 275 |
| Flame spread index and smoke developed index | Insulation, exterior facings and coatings shall be tested separately to ASTM E 84 or UL 723 and shall have a flame spread index of ≤25 and a smoke developed index of ≤450. (aluminium composite panels of ≤6.4 mm are permitted to be tested as an assembly) |
| Wall assembly flammability | The complete wall assembly must be tested and comply with NFPA 285 full-scale façade test |
| Potential heat content | Potential heat of foam plastic shall be determined applying NFPA 259. The potential heat of the foamed plastic in the installed walls shall not exceed that of the material tested in the full-scale façade test |
| Ignition characteristics | Exterior wall shall not produce sustained flaming when tested to NFPA 268 (ignitability of exterior walls using radiant heat). This requirement does not apply when the assembly is protected on the outside facing with complying facings such as 13 mm gypsum board, 9.5 mm glass reinforced concrete, 22 mm Portland cement plaster, 0.48 mm metal faced panels or 25 mm concrete or masonry |

Insulation other than foamed plastic, including vapour barriers and reflective foil insulation, must comply with the following requirements when tested to ASTM E 84 or UL 723 (NFPA 5000 Section 8.16):

- Concealed insulation—flame spread index of ≤75 and a smoke developed index of ≤450.
- Exposed insulation—flame spread index of ≤25 and a smoke developed index of ≤450.

Light transmitting plastic for exterior wall assemblies must comply with the following (NFPA 5000 Section 48.7)

- Self ignition temperature ≥343 °C (tested to ASTM D 1929 standard test method for determining ignition temperature of plastics);

- Smoke developed index of ≤450 (ASTM E 84 or UL 723);
- Be either CC1 (burn length ≤25 mm and self extinguishment) or CC2 (burning rate of ≤64 mm/min) when tested to ASTM D 635.

The CC1 or CC2 result impacts on the maximum area of plastic wall panels permitted and the minimum separation requirements.

Metal composite materials (MCM) must meet the following requirements (NFPA 5000 Section 37.4)

**Table B13** Metal composite material requirements

| Type I, Type II, Type III and Type IV buildings | |
|---|---|
| Height | Requirement |
| Installed to a maximum height of 12 m | Must either: |
| | • Cover <10 % or exterior wall area where the horizontal separation from the boundary is ≤1525 mm, or |
| | • Flame spread index of ≤75 and a smoke developed index of ≤450 (ASTM E 84 or UL 723) |
| Installed to a maximum height of 15 m | • Continuous areas of panels must not exceed 27.8 m² and must be separated from other continuous areas of panels by at least 1220 mm; and |
| | • Have a self ignition temperature ≥343 °C (tested to ASTM D 1929 standard test method for determining ignition temperature of plastics); and |
| | • Flame spread index of ≤75 and a smoke developed index of ≤450 (ASTM E 84 or UL 723) |
| Any height | • Compliance with NFPA 285 full scale façade test, And |
| | • Flame spread index of ≤25 and a smoke developed index of ≤450 (ASTM E 84 or UL 723) |
| Type V building | |
| Requirement for any height | Flame spread index of ≤75 and a smoke developed index of ≤450 (ASTM E 84 or UL 723) |

EIFS must be specified and installed in accordance with EIMA 99A (NFPA 5000 Section 37.5).

## Fire Stop Barriers

Internal gaps (e.g. between compartment floors the inside face of a wall such as a curtain wall) must be fire stopped with an approved material having a fire resistance at least equivalent to the compartment.

Use of fire stop barriers imbedded in EIFS or internal cavities of exterior wall systems are not specifically stated but would typically be required for compliance with the full scale façade fire test and EIFS Standards/guidelines specified.

**Separation Between Buildings**

The critical distance for separation of buildings from the boundary in terms of pro-tection of openings is 3 m. No unprotected openings are permitted at a separation distance of 3 m or less. At greater than 3 m the percentage of unprotected opening area permitted for external walls gradually increases to 100 % at a separation dis-tances of >10 m for most building types and >30 m for industrial and storage type buildings with ordinary and high hazard contents.

**Separation of Vertical Openings**

For buildings more than four stories in height which are not sprinkler protected openings must be separated from openings in the storey above by (NFPA 5000 Section 37.1.4) either:

- Protection of openings Sect 7.3, or
- A 915 mm spandrel with 1 h fire resistance
- A 760 mm horizontally projecting barrier with 1 h fire resistance.

**Sprinkler Protection**

Typical thresholds above which sprinkler systems are required in NFPA 5000, *Building Construction and Safety Code*, 2012 Edition include:

- Mercantile: Over 12,000 ft² (1115 m²) in gross fire area or three or more stories in height
- High-Rise: All buildings over 75 ft (22.9 m) in height
- Residential Apartments: All buildings except those in which each unit has indi-vidual exit discharge to the street

# UAE Fire and Safety Code

The 2011 version of the Fire and Life Safety Code of practice did not state any spe-cific requirements for combustible exterior wall systems. In response to a spate of fire incidents (primarily involving metal composite materials), Annexure A.1.21 of the UAE fire & life safety code was released which provides specific requirements for reaction to fire of exterior wall cladding and passive fire stopping.

## Requirements for Exterior Wall Materials

UAE Code Annexure A.1.21 states the following requirements for reaction to fire for combustible exterior wall systems to be tested as complete assemblies.

**Table B14**  UAE Code Annexure A.1.21 requirements for reaction to fire for combustible exterior wall systems

| Building types | Requirements |
|---|---|
| Mid rise (15–23 m high) or | • Class A when tested to ASTM E-84 (flame spread ≤25 and smoke development ≤450) |
| High Rise (>23 m high) or | • Class 1 or A1 when tested to FM 4880 |
| Low rise (<15 m high) having a horizontal separation of less than 3 m from boundary | • Class B1 or A2 when tested as per DIN 4102 and EN 13501-1 or ISO 9705 |
| | • BS 8414 Parts 1 or 2 as appropriate and classified in accordance with BR135 |
| | • 'Non Combustible' when tested to ASTM E 136 OR other equivalent test standards |
| Low rise (<15 m high) having a horizontal separation of 3 m or more from boundary | • Class B or Class II rating when tested as per NFPA 255 or ASTM E 84 or UL 723 (flame spread ≤75 and smoke development ≤450) |
| | • Class 0 when tested as per BS 476 part 6 and 7 |
| | • Class B2 when tested as per DIN 4102 |
| | • Class B as per EN 13501-1 |
| | • 'Equivalent of flame spread of less than 50' when tested to other equivalent test standards |

The document does not clearly state if wall systems for mid/high rise buildings are to be:

1. Only tested to one of the tests listed (either small scale or full scale façade test), or
2. Test to at least of the listed small scale tests AND the full scale test.

Comments from Exova Warringtonfire indicate that option 2 is the intended test requirement.

In addition to the above:

• For metal composite materials used as exterior walls, minimum exterior skin (0.5 mm and interior skin (0.25 mm) thicknesses and maximum panel thicknesses 0f 6.3 mm are required
• EIFS are to be in accordance with ANSI/EIMA 99-A, ASTM 1397 and ETAG 004. However it is not clear if compliance with all or only one of these standards/ guidelines is required.

**Fire Stop Barriers**

Internal gaps (e.g. between compartment floors the inside face of a wall such as a curtain wall) must be fire stopped with an approved material having a fire resistance at least equivalent to the compartment.

Use of fire stop barriers imbedded in EIFS or internal cavities of exterior wall systems are not specifically stated but would typically be required for compliance with the full scale façade fire test and EIFS Standards/guidelines specified in Annexure A.1.21.

**Separation Between Buildings**

The critical distance for separation of buildings from the boundary in terms of protection of openings is 3 m. No unprotected openings are permitted at a separation distance of 3 m or less. At greater than 3 m the percentage of unprotected opening area permitted for external walls gradually increases to 100 % at a separation distances of >10 m for most building types and >30 m for industrial and storage type buildings with ordinary and high hazard contents.

**Separation of Vertical Openings**

UAE Code Annexure A.1.21 states openings must be separated from openings in the storey above by either:

- A 915 mm spandrel with 1 h fire resistance
- A 760 mm horizontally projecting barrier with 1 h fire resistance.

No dispensation for sprinkler protected buildings is stated (however it is expected to be likely based on current design of UAE high rise buildings).

**Sprinkler Protection**

Sprinklers are required for assembly buildings, day care, healthcare, correctional, hotels/dormitory and residential board car buildings of nay height.

Sprinklers are required for educational. Mercantile, industrial and staff/labour accommodation >3 stories or 15 m high

Sprinklers are required for residential/apartments and business/office buildings >23 m high.

Sprinklers are also required when maximum compartment sizes are exceeded or fire resistance levels are reduced.

# Singapore Civil Defence Force Fire Code

The Singapore Civil Defence Force Fire Code is a performance based code which permits fire engineering Alternative Solutions.

## Requirements for Exterior Wall Materials

Requirements for cladding on external walls is stated in Fire Code Section 3.5.4.

**Table B15**  Singapore reaction to fire requirements for exterior wall cladding

| Building height (m) | Distance from relevant boundary (m) | Reaction to fire requirements |
|---|---|---|
| Any height | <1.0 | Cladding must be class 0. Class 0 is defined as either: |
| | | • Non-combustible |
| | | • Flame spread index ≤12 and sub index not exceeding 6 when tested to BS 476 Part 6 |
| >15 | ≥1.0 | • Cladding above 15 m must be class 0 |
| | | • Cladding below 15 m may be either timber ≥9 mm thick or a material tested to BS 476 Part 6 having a spread of flame index ≤20 |
| <15 | ≥1.0 | No requirements |

## Fire Stop Barriers

Internal gaps (e.g. between compartment floors the inside face of a wall such as a curtain wall) must be fire stopped with an approved material having a fire resistance at least equivalent to the compartment (Fire Code Section 3.7.3).

Fire stopping of any cavity must be provided to close the edges of a cavity around openings in an external wall. Fire stoping of cavity if required at max 20 m intervals the cavity is constructed of Class 0 materials, or max 8 m intervals if lower class materials are used.

Use of fire stop barriers imbedded in EIFS is not specifically stated.

## Separation Between Buildings

The critical distance for separation of buildings from the boundary in terms of protection of openings is 1 m. No unprotected openings are permitted at a separation distance of 1 m or less. At greater than 1 m the percentage of unprotected opening area permitted for external walls gradually increases to 100 % dependant on building height.

**Separation of Vertical Openings**

Review of the fire code did not identify any requirements for separation of vertical separation of openings in floors above and below.

**Sprinkler Protection**

For all building types except residential buildings sprinklers are required for buildings exceeding 24 m in height. Sprinklers are also required when compartment size limits are exceeded.

# Malaysian Uniform Building By-Laws

The Uniform Building By-Laws 1984 state the prescriptive fire safety requirements for buildings in Malaysia. Fire engineering alternative solutions are generally only permitted relating to fire compartment size, smoke hazard management and exit distances and locations. Any alternative solutions must be approved by BOMBA (Malaysian fire brigade).

**Requirements for Exterior Wall Materials**

Requirements for cladding on external walls is stated in By-Law 144.

**Table B16** Malaysian reaction to fire requirements for exterior wall cladding

| Building height (m) | Distance from relevant boundary (m) | Reaction to fire requirements |
|---|---|---|
| Any height | <1.2 | Cladding must be class 0. Class 0 is defined as either: |
| | | • Non-combustible |
| | | • Flame spread index ≤12 and sub index not exceeding 6 when tested to BS 476 Part 6 |
| >18 | ≥1.2 | • Cladding above 18 m must be class 0 |
| | | • Cladding below 18 m may be either timber ≥10 mm thick or a material tested to BS 476 Part 6 having a spread of flame index ≤20 |
| <18 | ≥1.2 | No requirements |

**Fire Stop Barriers**

Internal gaps (e.g. between compartment floors the inside face of a wall such as a curtain wall) must be fire stopped with an approved material having a fire resistance at least equivalent to the compartment (By-Law 161).

Fire stopping of any cavity with a surface of combustible material exposed within the cavity which is of a class lower than Class 0 must be fires stopped so that the cavity does not exceed 7.625 m in a single direction or 23.225 m² in area.

Use of fire stop barriers imbedded in EIFS is not specifically stated.

### Separation Between Buildings

The critical distance for separation of buildings from the boundary in terms of protection of openings is 1 m. No unprotected openings are permitted at a separation distance of 1 m or less. At greater than 1 m the percentage of unprotected opening area permitted for external walls gradually increases to 100 % dependant on building height.

### Separation of Vertical Openings

By-Law 149 states openings must be separated from openings in the storey above by flame barriers being either:

*   A 915 mm spandrel
*   A 760 mm horizontally projecting barrier with.

A fire rating for the flame barriers is not specified and no dispensation for sprinkler protected buildings is stated.

### Sprinkler Protection

Sprinklers are required for assembly buildings any height.
Sprinklers are required for healthcare buildings and hotels exceeding 15 m high
Sprinklers are required for Apartments of ten storeys or more
Sprinklers are required for Offices >30 m
Sprinklers are also required when compartment size limits are exceeded (this is the main trigger for sprinklers for shops, industrial and storage buildings.

# National Fire Code of Canada

It is understood that the CAN/ULC S134 full scale façade test is adopted by the Canadian Fire Code. A copy of the fire code was not obtained for this project so detailed review of small scale reaction to fire test requirements for exterior walls, separation of buildings and openings and sprinkler protection requirements was not undertaken.

Exterior non-load bearing wall assemblies containing combustible components are permitted provided that:

- the building is not more than three storeys unsprinklered or sprinklered if more than three storeys; and
- the interior surfaces of the wall assembly are protected with a thermal barrier; and
- the wall assembly is subjected to the full scale facade test method of CAN/ULC S134 and flaming does not spread more than 5 m above the opening during or following the test, and the heat flux during flame exposure on the wall assembly is not more than 35 kW/m$^2$ measured 3.5 m above the opening

# Appendix C: Large and Intermediate Scale Façade Fire Test Summary Table

© Fire Protection Research Foundation 2015

N. White, M. Delichatsios, *Fire Hazards of Exterior Wall Assemblies Containing Combustible Components*, SpringerBriefs in Fire, DOI 10.1007/978-1-4939-2898-9

**Table C1** Large and intermediate scale façade fire test summary

| Test Standard | | Full-scale façade tests | | | | |
|---|---|---|---|---|---|---|
| | | ISO 13785 Part 1:2002 | BS 8414 part 1 | BS 8414 part 2 | DIN 4102-20 (Draft) | NFPA 285 |
| Country used | | International | UK | UK | Germany | USA |
| Test Scenario | | Flames emerging from a flashover compartment fire via a window | Flames emerging from a flashover compartment fire via a window | Flames emerging from a flashover compartment fire via a window | Flames emerging from a flashover compartment fire via a window | Flames emerging from a flashover compartment fire via a window |
| Summary geometry of test rig | Number of walls | Two walls in re-entrant corner "L" arrangement | Two walls in re-entrant corner "L" arrangement | Two walls in re-entrant corner "L" arrangement | Two walls in re-entrant corner "L" arrangement | One wall |
| | Number of openings | 1 (fire source opening) | 1 (fire compartment opening) | 1 (fire compartment opening) | 1 (fire compartment opening) | 1 (fire compartment opening) |
| Fire source | Standard source | Series of large perforated pipe propane burners. Total peak output 120 g/s (5.5 MW) within standard fire enclosure | Timber crib 1.5 m wide × 1 m deep × 1 m high. Nominal heat output of 4500 MJ over 30 min. Peak HRR = 3 ± 0.5 MW. Crib located on platform 400 mm above base of test rig | Same as BS 8414 part 1 | 320 kW constant HRR linear gas burner located approx. 200 mm below soffit of opening | Rectangular pipe gas burner in fire compartment (room burner) 1.52 m long pipe gas burner near opening soffit (window burner) Room burner increases from 690 to 900 kW over 30 min test period Window burner ignited 5 min after room burner and increases from 160 to 400 kW over remaining 25 min test period |
| | Alternative source | Liquid pool fires or 16×25 kg timber cribs distributed on floor of standard fire enclosure | Permitted but must achieve calibration requirements | Same as BS 8414 part 1 | 25 kg timber crib, 0.5 m×0.5 m×0.48 m, using 40 mm×40 mm softwood sticks | Not specified or permitted by standard |
| Fire exposure | Calibrated heat flux exposure (with non-combustible wall) | 55 ± 5 kW/m² at a height of 0.6 m above opening 35 ± 5 kW/m² at a height of 1.6 m above opening | Mean within range of 45–95 kW/m² at height of 1 m above opening over continuous 20 min period. Typical steady state mean of 75 kW/m² at height of 1 m above opening within this period | Same as BS 8414 part 1 | 60 kW/m² at 0.5 m above opening 35 kW/m² at 1.0 m above opening 25 kW/m² at 1.5 m above opening | 38 ± 8 kW/m² at 0.6 m above opening during peak fire source period 25–30 min 40 ± 8 kW/m² at 0.9 m above opening during peak fire source period 25–30 min 34 ± 7 kW/m² at 1.2 m above opening during peak fire source period 25–30 min |
| | Calibrated temperature exposure (with non-combustible wall) | >800 °C at 50 mm above opening | >600 °C above ambient within fire compartment >500 °C above ambient on exterior of non-combustible wall 2.5 m above opening | Same as BS 8414 part 1 | Maximum temp. of 780–800 °C on exterior of non-combustible wall 1 m above opening soffit | Average 712 °C on exterior of non-combustible wall 0.91 m above opening Average 543 °C on exterior of non-combustible wall 1.83 m above opening |
| | Maximum height of flames extending above opening for non-combustible wall | – | Approx. 2.5 m | Same as BS 8414 part 1 | Approx. 2.5 m | Approx. 2.0 m |
| | Duration | 23–27 min. 4–6 min growth phase, approx. 15 min steady state phase, 4–6 min decay phase | 30 min (approx. 7 min growth phase) | Same as BS 8414 part 1 | 20 min (gas burner) 30 min (crib) | 30 min |

| | | Intermediate scale façade tests | | | | |
|---|---|---|---|---|---|---|
| SP FIRE 105 | CAN/ULC S134 | FM 25 ft high corner test | FM 50 ft high corner test | ISO 13785 Part 2:2002 | ASTM vertical channel test | BRANZ vertical channel test |
| Sweden | Canada | US/International | US/International | | Canada | New Zealand |
| Flames emerging from a flashover compartment fire via a window | | External (or internal) pellet fire located directly against the base of a re-entrant wall corner | External (or internal) pellet fire located directly against the base of a re-entrant wall corner | Flames at base of small section of façade | Flames emerging from a flashover compartment fire via a window | Flames emerging from a flashover compartment fire via a window |
| One wall | One wall | Two walls in re-entrant corner "L" arrangement. Ceiling over top of walls | Two walls in re-entrant corner "L" arrangement. Ceiling over top of walls | Two walls in re-entrant corner "L" arrangement | One wall | One wall |
| 2 (fire compartment opening and fictitious window above) | 1 (fire compartment opening) | 0 | 0 | 0 | 1 (fire compartment opening at base of test wall) | 1 (fire compartment opening at base of test wall) |
| Heptane fuel tray, 0.5 m wide × 2.0 m long × 0.1 m high. Filled with 60 l Heptane. Approx. 2.5 MW peak | Four 3.8 m long linear propane burners. Total output 120 g/s propane (5.5 MW) | 340 ± 4.5 kg crib constructed of 1.065 m 1.065 m oak pallets, max height 1.5 m. Located in corner 305 mm from each wall. Ignited using 0.24 L gasoline at crib base | Same as FM 25 ft test | Constant 100 kW linear propane burner 1.2 m long × 0.1 m wide located 0.25 m below bottom edge of main wall | Two propane gas burners 1.16 MW typical output | Same as ASTM vertical channel test |
| Permitted but must achieve calibration requirements | Wood cribs of kiln dried pine with total mass of 675 kg | Not specified or permitted by standard | Not specified or permitted by standard | Not specified or permitted by standard | Not specified or permitted by standard | Not specified or permitted by standard |
| 15 kW/m² at 4.8 m above opening during at least 7 min of the test | 45 ± 5 kW/m² at 0.5 m above opening averaged over 15 min steady state period | Not specified | Not specified | Not specified | 50 ± 5 kW/m² at 0.5 m above the opening averaged over 20 min steady burner output | Same as ASTM vertical channel test |
| 35 kW/m² at 4.8 m above opening during at least 1.5 min of the test | 27 ± 3 kW/m² at 1.5 m above opening averaged over 15 min steady state period | | | | 27 ± 3 kW/m² at 1.5 m above the opening averaged over 20 min steady burner output | |
| <75 kW/m² at 4.8 m above opening at all times | | | | | | |
| Not specified | – | Not specified | Not specified | Not specified | Not specified | Not specified |
| – | Approx. 2.0 m | – | – | Approx. 0.2 m | – | – |
| Approx. 15 min | 25 min. 5 min growth phase, 15 min steady state phase, 5 min decay phase | approx. 15 min | same as FM 25 ft test | 30 min | 20 min | Same as ASTM vertical channel test |

(continued)

**Table C1**  (continued)

| Test Standard | | Full-scale façade tests | | | | |
|---|---|---|---|---|---|---|
| | | ISO 13785 Part 1:2002 | BS 8414 part 1 | BS 8414 part 2 | DIN 4102-20 (Draft) | NFPA 285 |
| Country used | | International | UK | UK | Germany | USA |
| Detailed geometry of test rig | Total height of apparatus | ≥5.7 m | ≥8.0 m | Same as BS 8414 part 1 | ≥5.5 m | ≥5.33 m |
| | Height of test wall above fire compartment opening | ≥4.0 m | ≥6.0 m | Same as BS 8414 part 1 | ≥4.5 m | ≥4.52 m |
| | Width of main test wall | ≥3.0 m | ≥2.5 m | Same as BS 8414 part 1 | ≥2.0 m (using gas burner)  ≥1.8 m (using crib) | ≥4.1 m |
| | Width of wing test wall | ≥1.2 m | ≥1.5 m | Same as BS 8414 part 1 | ≥1.4 m (using gas burner)  ≥1.2 m (using crib) | N/A |
| Detailed geometry of test rig (continued) | Height of fire compartment opening above bottom of test wall | 0.5 m | 0 m | Same as BS 8414 part 1 | 0 m | 0.76 m |
| | Height of fire compartment opening | 1.2 m | 2 m | Same as BS 8414 part 1 | 1 m | 0.76 m |
| | Width of fire compartment opening | 2 m | 2 m | Same as BS 8414 part 1 | 1 m | 1.98 m |
| | Horizontal distance of opening from wing wall | 50 mm | 250 mm | Same as BS 8414 part 1 | 0 mm | N/A |
| | Fire compartment dimensions | 4 m wide × 4 m deep × 2 m high with 0.3 m deep soffit across opening  Alternative sizes permitted in range of 20–30 m³ | 2 m wide × 2 m high (depth not specified) | Same as BS 8414 part 1 | 1 m wide × 1 m high | 3 m wide × 3 m deep × 2 m high |
| Test wall substrate | | Details of substrate or supporting frame not specified by standard | Masonry | Steel frame (open) to support complete test wall assembly | Aerated concrete | Steel frame and concrete floor slabs (open) to support complete test wall assembly |
| Test measurements | Heat flux at surface test wall | 0.6 m, 1.6 m and 3.6 m above opening | Not required | Same as BS 8414 part 1 | – | Not required |
| | Temperatures | Wall exterior and intermediate layers/ Cavities immediately above window and at 4 m above window | Wall exterior at 2.5 and 5.0 m above opening  Intermediate layers and cavities at 5.0 m above opening | Same as BS 8414 part 1 | Wall exterior and intermediate layers/ Cavities at 3.5 m above opening | Wall exterior and intermediate layers/ cavities at 305 mm intervals vertically above opening  At rear of test wall within second storey room enclosure |

| | | | | Intermediate scale façade tests | | |
|---|---|---|---|---|---|---|
| SP FIRE 105 | CAN/ULC S134 | FM 25 ft high corner test | FM 50 ft high corner test | ISO 13785 Part 2:2002 | ASTM vertical channel test | BRANZ vertical channel test |
| Sweden | Canada | US/International | US/International | | Canada | New Zealand |
| 6.71 m | 10.0 m | 7.6 m | 15.2 m | 2.8 m | 9.4 m | 7.1 m |
| 6.0 m | 7.25 m | N/A | N/A | 2.4 m (no opening) | 7.32 m | 5 m |
| 4.0 m | 5.0 m | 15.7 m (specimen installed to full width over top 3.8 m and to 6 m out from corner for bottom 3.8 m) | 6.2 m | 1.2 m | 0.8 m | Same as ASTM vertical channel test |
| N/A | N/A | 11.96 m (specimen installed to full width over top 3.8 m and to 6 m out from corner for bottom 3.8 m) | 6.2 m | 0.6 m | Non-combustible 0.5 m wide wing wall on both sides of test wall to form vertical channel | Same as ASTM vertical channel test |
| 0 m | 1.5 m | N/A | N/A | N/A | 0 m | Same as ASTM vertical channel test |
| 0.71 m | 1.37 m | N/A | N/A | N/A | 0.63 m | Same as ASTM vertical channel test |
| 3.0 m | 2.6 m | N/A | N/A | N/A | 0.8 m | Same as ASTM vertical channel test |
| N/A | N/A | N/A | N/A | N/A | N/A | Same as ASTM vertical channel test |
| 3.0 m wide × 1.6 m deep × 1.3 m high | 5.95 m wide × 4.4 m deep × 2.75 m high | N/A | N/A | N/A | 0.8 m wide × 1.5 m deep × 1.9 m high | Same as ASTM vertical channel test |
| Steel frame (open) to support complete test wall assemblies<br><br>Light weight concrete substrate to support claddings which require such a substrate | Concrete | Steel frame (open) to support complete test wall assembly | Steel frame (open) to support complete test wall assembly | Non-combustible board (thickness 12 m, Nominal Density 750 kg/m³) | Steel frame (open) to support complete test wall assembly | Same as ASTM vertical channel test |
| 2.1 m above opening (centre of ficticiuos first storey window) | 3.5 m above opening | Not required | Not required | vertical intervals of 0.5 m on the centre of both test wall surfaces | 3.5 m above opening | Same as ASTM vertical channel test |
| Minimum 2 thermocouples measuring gas temperatures at top of wall on underside of 500 mm non combustible eave | Within fire enclosure and at opening 0.15 m below soffit<br><br>Wall exterior and intermediate layers/cavities at vertical intervals of 1 m starting from 1.5 m above opening<br><br>Gas temperatures 0.6 m in front of the top of the test wall | Exterior of exposed side of test walls on a 2.5 m grid spacing | Near intersection of top of walls and ceiling, both at the wall corner and 4.6 m out from the wall corner | Centre top of main test wall | Wall exterior and intermediate layers/cavities at vertical intervals of 1 m starting from 1.5 m above opening | Same as ASTM vertical channel test |

(continued)

**Table C1** (continued)

| | | Full-scale façade tests | | | | |
|---|---|---|---|---|---|---|
| Test Standard | | ISO 13785 Part 1:2002 | BS 8414 part 1 | BS 8414 part 2 | DIN 4102-20 (Draft) | NFPA 285 |
| Country used | | International | UK | UK | Germany | USA |
| Performance criteria | External fire spread | Reported—Criteria not specified by standard | Fire spread start time = time external temp at level 1 (2.5 m above opening) exceeds 200 °C above ambient | Same as BS 8414 part 1 | • No burned damaged (excluding melting or sintering) ≥3.5 m above opening | • Wall exterior temp must not exceed 538 °C at 3.05 m above opening |
| | | | Level 2 external temp (5 m above opening) must not exceed 600 °C above ambient (over >30 s), within 15 min of fire spread start time | | • Temperatures on wall surface or within the wall layers/cavities must not exceed 500 °C ≥3.5 m above opening | • Exterior flames must not extend vertically more than 3.05 m above opening |
| | | | | | • No observed continuous flaming for more than 30s ≥3.5 m above opening | • Exterior flames must not extend horizontally more than 1.52 m from opening centreline |
| | | | | | • No flames to the top of the specimen at any time | • Flames must not occur horizontally beyond the intersection of the test wall and the side walls of the test rig |
| | Internal fire spread | | Level 2 internal temp (5 m above opening) must not exceed 600 °C above ambient (over >30 s), within 15 min of fire spread start time | Same as BS 8414 part 1 | • No burned damaged (excluding melting or sintering) ≥3.5 m above opening | • Fire spread horizontally and vertically within wall must not exceed designated internal wall cavity and insulation temp limits. Position of designated thermocouples and temp limits depends on type/thickness of insulation and whether or not an air gap cavity exists |
| | | | | Plus, Flaming (>60 s) must not occur on non-exposed side of the test wall at height of ≥0.5 m within 15 min of fire spread start time | • Temperatures within the wall layers/cavities must not exceed 500 °C ≥ 3.5 m above opening | • Temp at the rear of test wall in second storey test room must not exceed 278 °C above ambient |
| | | | | | | • Flames shall not occur in the second story test room |

| | | | | Intermediate scale façade tests | | |
|---|---|---|---|---|---|---|
| SP FIRE 105 | CAN/ULC S134 | FM 25 ft high corner test | FM 50 ft high corner test | ISO 13785 Part 2:2002 | ASTM vertical channel test | BRANZ vertical channel test |
| Sweden | Canada | US/International | US/International | | Canada | New Zealand |
| No fire spread (flame and damage) >4.2 m above opening (bottom of second storey ficticious window) | Flame spread distance less than 5 m above the opening soffit | The tested assembly shall not result in fire spread to the limits of the test structure as evidenced by flaming or material damage | Must meet requirements for 25 ft test | Reported—Criteria not specified by standard | Flame spread distance less than 5 m above the opening soffit | Same as ASTM vertical channel test |
| Temps at the eave must not exceed 500 °C for more than 2 min or 450 °C for more than 10 min | Heat flux 3.5 m above opening must be less than 35 kW/m$^2$ | | • For acceptance to maximum height use of 50 ft (15.2 m), tested assembly shall not result in fire spread to limits of test structure as evidenced by flaming or material damage. | | Heat flux 3.5 m above opening must be less than 35 kW/m$^2$ | |
| Additionally, for buildings >8 stories high or hospitals of any height, Heat flux at 2.1 m above opening must not exceed 80 kW/m$^2$ | | | • For acceptance to unlimited height use tested assembly shall not result in fire spread to the limits of the test structure or to the intersection of the top of the wall and the ceiling as evidenced by flaming or material damage | | | |
| No fire spread (flame and damage) >4.2 m above opening (bottom of second storey ficticious window) | Flame spread distance less than 5 m above the opening soffit | The tested assembly shall not result in fire spread to the limits of the test structure as evidenced by flaming or material damage | Must meet requirements for 25 ft test | Reported—criteria not specified by standard | Flame spread distance less than 5 m above the opening soffit | |
| | | | • For acceptance to maximum height use of 50 ft (15.2 m), tested assembly shall not result in fire spread to limits of test structure as evidenced by flaming or material damage | | Heat flux 3.5 m above opening must be less than 35 kW/m$^2$ | |
| | | | • For acceptance to unlimited height use tested assembly shall not result in fire spread to the limits of the test structure or to the intersection of the top of the wall and the ceiling as evidenced by flaming or material damage | | | |

(continued)

**Table C1** (continued)

| Test Standard | | Full-scale façade tests | | | | |
|---|---|---|---|---|---|---|
| | | ISO 13785 Part 1:2002 | BS 8414 part 1 | BS 8414 part 2 | DIN 4102-20 (Draft) | NFPA 285 |
| Country used | | International | UK | UK | Germany | USA |
| | Burning debris and droplets | Reported—Criteria not specified by standard | Reported—Criteria not specified | Same as BS 8414 part 1 | Falling burning droplets and burning and non-burning debris and lateral flame spread must cease with 90 s after burners off | Reported—Criteria not specified by standard |
| | Mechanical behaviour | Reported—Criteria not specified by standard | Reported—Criteria not specified | Same as BS 8414 part 1 | Reported—Criteria not specified | Reported—Criteria not specified by standard |
| Comments | | | | | | |

*Note:* Please note that in Table 1 of Appendix C, the columns are split-up and placed in facing pages.

| SP FIRE 105 | CAN/ULC S134 | FM 25 ft high corner test | FM 50 ft high corner test | Intermediate scale façade tests | | |
|---|---|---|---|---|---|---|
| | | | | ISO 13785 Part 2:2002 | ASTM vertical channel test | BRANZ vertical channel test |
| Sweden | Canada | US/International | US/International | | Canada | New Zealand |
| Reported—Criteria not specified by standard | Reported—Criteria not specified by standard | Reported—Criteria not specified by standard | Reported—Criteria not specified by standard | Reported—Criteria not specified by standard | Reported—Criteria not specified by standard | Reported—Criteria not specified by standard |
| No large pieces may fall from the façade | Reported—Criteria not specified by standard | Reported—Criteria not specified by standard | Reported—Criteria not specified by standard | Reported—Criteria not specified by standard | Reported—Criteria not specified by standard | Reported—Criteria not specified by standard |
| Includes two fictitious window details in test wall and level 1 and level 2 blacked at rear with non combustible lining | | Mostly only used for insulated sandwich panel | Mostly only used for insulated sandwich panel | Intended as reduced cost screening and product development test for ISO 13785-1 | Developed as intermediate test for CAN/ULC S134 | Same as ASTM vertical channel test with reduced wall height |

# Appendix D: Existing Research References

The following is an extended list of related research literature. It is not practical to summarise all of this research, however this list is provided for further reading.

[1]. MATEC Web Conferences, Proceedings 1st International Seminar for Fire Safety, 14-15 November 2013, Paris, France, where many other references can be found. http://www.matecconferences.org/index.php?option=com_toc&url=/articles/matecconf/abs/2013/07/contents/contents.html

[2]. Morris, J.: 'The First Interstate Bank fire—what went wrong?', Fire Prevention, No.226, January/February 1990, p 20-26. FPA Casebook of Fires, High Rise Offices, Fire Prevention No.248, April 1992, p 36-38.

[3]. Klem, T. J.: 'High-rise fire claims three Philadelphia fire fighters', NFPA Journal.

[4]. Ashton, L. A., Malhotra, H. L.: 'External walls of buildings—Part 1, The protection of openings against spread of fire from storey to storey', F. R. Note No.436, July 1960, 8pp.

[5]. Belles, D. W., Beitel, J. J.: 'Between the cracks...How fire spreads from floor to floor in a building with aluminium curtain walls', Fire Journal, May/June 1988, p 75-84.

[6]. Morris, B., Jackman, L. A.: 'Fire spread in multi-storey buildings with glazed curtain wall facades", LPR 11, 1999, 56pp.

[7]. Jackman, L., Finegan, M., Morris, B.: 'Interim report on Test 12, multi-storey test', BPSG/98/20, December 1997, 10pp.

[8]. Arvidson, M., An Initial Evaluation of Different Residential Sprinklers using HR Measurements, SP Rapport 2000:18, The Technical Research Institute of Sweden (SP), Borås, 2000.

[9]. British Standards Institution (BSI). Fire safety engineering principles for the design of buildings. BS 7974:2001, London, 2001. CAENZ, see New Zealand Centre for Advanced Engineering.

[10]. European Standard, Eurocode 1: Actions on structures—Part 1-2: General actions—Actions on structures exposed to fire, EN 1991-1-2, November 2002.

[11]. European Standard, Fire classification of construction products and building elements—Part 2: Classification using data from fire resistance tests, excluding ventilation services, EN 13501-2, 2007.

[12]. European Standard, Fixed firefighting systems. Automatic sprinkler systems. Design, installation and maintenance, EN 12845, 2009.

[13]. Fire Code Reform Centre, Fire performance of wall and ceiling lining materials, CRC Project—Stage A, Fire Performance of Materials, Project Report FCRC—PR 98-02, Fire Code Reform Research Program, FCRC, Sydney, 1998.

© Fire Protection Research Foundation 2015
N. White, M. Delichatsios, *Fire Hazards of Exterior Wall Assemblies Containing Combustible Components*, SpringerBriefs in Fire,
DOI 10.1007/978-1-4939-2898-9

[14].   Frantzich H. Tid för utrymning vid brand, SRV rapport P21-365/01. Statens räddningsverk, Karlstad, 2001.

[15].   Hall, J.R., U.S. Experience with Sprinklers and other Fire Extinguishing Equipment, Fire Analysis and Research Division National Fire Protection Association, February 2010.

[16].   Höglander K., Sundström B., Design fires for pre-flashover fires, SP Report 1997:36,Swedish National Testing and Research Institute, Borås, 1997.

[17].   Lougheed, G.D., Expected size of shielded fires in sprinklered office buildings, ASHRAE Transactions, Vol. 103, Pt. 1, American Society of Heating, Refrigerating and Air-Conditioning Engineers, Inc., 1997.

[18].   Lundin, J. Safety in Case of Fire—The Effects of Changing Regulations, Dept. of Fire Safety Engineering, Lund University, Lund, 2005. Case studies on the verification of fire safety design in sprinklered buildings.

[19].   McGrattan, K.B., Hostikka, S., Floyd, J., Fire Dynamic Simulator (Version 5)—User's Guide, NIST Special Publication 1019-5, National Institutes of Standards and Technology, USA, 2010.

[20].   New Zealand Centre for Advanced Engineering (CAENZ), Fire Engineering Design Guide, 3rd edition, Christchurch, 2008.

[21].   National Fire Protection Association (NFPA), Standard for the installation of sprinkler systems in residential occupancies up to and including four stories in height, NFPA 13R, Quincy, 2010.

[22].   Nordic Committee on Building Regulations (NKB), Performance Requirements for Fire Safety and Technical Guide for Verification by Calculation, Report nr. 1994:07 E, 1994.

[23].   Nystedt, F., Deaths in Residential Fires—an Analysis of Appropriate Fire Safety Measures, report 1026, Department of Fire Safety engineering, Lund University, 2003.

[24].   Nystedt, F., Verifying Fire Safety Design in Sprinklered Buildings, report 3150, Department of Fire Safety Engineering and Systems Safety, Lund University, 2011.

[25].   Nystedt, F., Frantzich, H., Kvalitetsmanual för brandtekniska analyser vid svenska kärntekniska anläggningar, report 3160, Department of Fire Safety Engineering and Systems Safety, Lund University, 2011. (In Swedish).

[26].   Platt, D.G., Fire resistance of barriers in modelling fire spread, Fire Safety Journal 22, pp.399-407, Elsevier Science Limited, 1994.

[27].   Society of Fire Protection Engineers (SFPE), SFPE Engineering Guide to Performance-Based Fire Protection, 2nd edition, Bethesda, 2007.

[28].   Suzuki, T., Sekizawa, A., Yamada, T., Yanai, E., Satoh, H., Kurioka, H., Kimura, Y., An experimental study of ejected flames of a high-rise buildings, Technical report, National Research Institute of Fire and Disaster, Japan, 2001. p. 363–73.

[29].   Trätek, Brandsäkra trähus (version 2)—nordisk kunskapsöversikt och vägledning, Publ nr0210034, Stockholm, 2002. (In Swedish)

[30].   Suzuki, T., Sekizawa, A., Yamada, T., Yanai, E., Satoh, H., Kurioka, H. and Kimura, Y., "An Experimental Study of ejected Flames of High Rise Buildings—Effects of

[31].   Depth of Balcony on Ejected Flame, "Proceedings of the Fourth Asia-Oceania Symposium on Fire Science and Technology", Asia-Oceania Association for Fire Science and Technology, Tokyo, Japan, 2000, pp.363-373

[32].   Thomas, P.H. and Law, M., "The Projection of Flames from Burning Buildings on Fire", Fire Prevention Science and Technology No.10: 19-26 (1974)

[33].   Jansson, R. and Onnermark, B., "Flame Heights Outside Windows", National Defense Research Inst., Stockholm, Sweden FOA Report, C20445-AC, 1982

[34].   Hasemi, Y. and Izumu, J., "Leading Mechanism and General Correlation of Flames from an Opening of Fire Enclosure", Summaries of Technical Papers of Annual Meeting Architectural Inst. of Japan, Fire safety section (in Japanese) 1988, pp.745-746

[35].   Hägglund, B., Jansson, R. and Onnermark, B., "Fire Development in Residential Room after Ignition from Nuclear Explosions", National Defense Research Inst., Stockholm, Sweden FOA Report, C2016-D6(A3), 1974

[36]. Sugawa, O. and Takahashi, W., "Flow Behavior of Ejected Fire Flame/Plume from an Opening Effected by External Side Wind", Fire Safety Science—Proceedings of the Fifth International Symposium on Fire Safety Science, International Association of Fire Safety Science, Melbourne, Australia, 1997, pp.249-260

[37]. Takanashi, K., Suzuki, T., Sekizawa, A., Yamada, T., Yanai, E., Satoh, H., Kurioka, H. and Kimura,Y., "Flame Configuration derived from Video Camera Picture—An Experimental Study of Ejected Flames of a High-rise Building Part II", Report of National Research Institute of Fire and Disaster No.91, Mitaka, JAPAN, 2001, pp.48-58

[38]. Bøhm, B. and Rasmussen, B.M., "The Development of a Small-scale Fire Compartment in order to Determine Thermal Exposure Inside and Outside Buildings", Fire Safety Journal, 12 : 103-108 (1987)

[39]. Suzuki, H. "Fire Safety Performance of Material and Products". Fire Handbook—Kasai Binran (3rd ed.), (in Japanese), Japan Association for Fire Science and Engineering, 1997, p.803

[40]. Fleischmann, C.N. and Parkes, A.R., "Effects of Ventilation on the Compartment enhanced Mass Loss Rate", Fire Safety Science—Proceedings of the Fifth International Symposium on Fire Safety Science, International Association of Fire Safety Science, Melbourne, Australia, 1997, pp. 415-426

[41]. Law, M. and O'Brien, T. (1989), Fire safety of bare external structural steel, The Steel Construction Institute, SCI Publication 009, U.K. ISBN: 0 86200 026 2.

[42]. BSI (2002), Eurocode 1: Actions on structures—Part 1-2: General actions—Actions on structures exposed to fire, British Standards, U.K. BS EN 1991-1-2:2002.

[43]. Balter, M. (2004), Earliest signs of human-controlled fire uncovered in Israel, Science 304 (5671), pp.663—665.

[44]. Plato (ca. 360 BC) Timaeus, translated by W.R.M. Lamb in Plato in Twelve Volumes (9), Cambridge, MA, Harvard University Press; London, William Heinemann Ltd.

[45]. Richardson, J.K., (Ed.) (2003), History of Fire Protection Engineering, National Fire Protection Association, Inc., Massachusetts, USA.

[46]. Emmons, H.W. (1984), The further history of fire science, Combustion Science and Technology 40, pp.167-174. Gordon and Breach Science Publishers.

[47]. Cote, A.E. (2008), History of fire protection engineering, Fire Protection Engineering (10 Year magazine), Fall of 2008, Society of Fire Protection Engineers, U.S.A., pp.28-36.

[48]. Beyler, C. (1999), Guest editorial: Professor Howard Emmons 1912-1998, Fire Technology 35 (1), pp.1-3.

[49]. Trouvé A. (2007), Challenges in CFD modeling of large-scale pool fires, Proceedings of Advanced Research Workshop on Fire Computer Modeling, (Ed.) J. A. Capote Abreu, GIDAI, Univ. Cantabria, Santander, Spain, pp.1-14.

[50]. BBC (2007), Skyscraper Fire Fighters (documentary), BBC Horizon, U.K. www.bbc.co.uk/sn/tvradio/programmes/horizon/broadband/tx/firegrid/ [accessed 16/09/2009]

[51]. McAllister, T., Barnett, J., Gross, J., Hamburger, R., Magnusson, J. (2002), World Trade Center building performance study: data collection, preliminary observations and recommendations, Chapter 1: Introduction, Technical Report FEMA 403, Federa Emergency Management Agency, Washington, D.C., U.S.A.

[52]. Usmani, A.S., Chung, Y.C. and Torero, J.L. (2003), How did the WTC towers collapse: anew theory, Fire Safety Journal 38 (6), pp.501-533.

[53]. Quintiere, J.G. (1989), Scaling applications in fire research, Fire Safety Journal 15 (1), pp.3-29.

[54]. Ma, T.G. and Quintiere, J.G. (2003), Numerical simulation of axi-symmetric fire plumes: accuracy and limitations, Fire Safety Journal 38 (5), pp.467-492.

[55]. Rein, G., Torero, J.L., Wolfram, J., Stern-Gottfried, J., Ryder, N.L., Desanghere, S., Lázaro, M., Mowrer, F., Coles, A., Joyeux, D., Alvear, D., Capote, J.A., Jowsey, A., Abecassis-Empis, C. and Reszka, P. (2009), Round-robin study of a priori modelling predictions of the Dalmarnock Fire Test One, Fire Safety Journal 44 (4), pp.590-602

[56]. Jahn, W., Rein, G. and Torero, J.L. (2007), Posteriori modelling of Fire Test One, Chapter 11, The Dalmarnock Fire Tests: Experiments and Modelling, (Eds.) Rein, G., Abecassis Empis, C. and Carvel, R., School of Engineering and Electronics, The University of Edinburgh, U.K. ISBN 978-0-9557497-0-4.

[57]. Fernandez-Pello, A.C. (1994), The solid phase, Chapter 2, Combustion Fundamentals of Fire, (Ed.) G. Cox, Academic Press, pp.31-100.

[58]. NFPA (2003), Fire Protection Handbook, 19th Edition, National Fire Protection Association, U.S.A.

[59]. BSI (2001), Application of fire safety engineering principles to the design of buildings—Code of Practice, British Standards, U.K. BS 7974:2001.

[60]. Cowlard, A., Jahn, W., Abecassis-Empis, C., Rein, G. and Torero, J.L. (2010), Sensor assisted fire fighting, Fire Technology, 46 (3), pp.719-741.

[61]. Upadhyay, R., Pringle, G., Beckett, G., Potter, S., Han, L., Welch, S., Usmani, A. And Torero, J. (2009), An architecture for an integrated fire emergency response system for the built environment, Fire Safety Science 9, pp.427-438.

[62]. Law, M. (1978), Fire safety of external building elements—The design approach, Engineering Journal, Second Quarter, pp.59-74, American Institute of Steel Construction (AISC), U.S.A.

[63]. Law, M. (1981), Designing fire safety for steel—recent work, ASCE Spring Convention, American Society of Civil Engineers, U.S.A.

[64]. Tonkelaar, E. den (2003), Prediction of the effect of breaking windows in a double skin facade as a result of fire, Eighth International IBPSA Conference, Eindhoven, Netherlands.

[65]. Colwell, S. and Martin, B. (2003), Fire performance of external thermal insulation for walls of multi-storey buildings, BR 135, Building Research Establishment (BRE), U.K. ISBN 1-86081-622-3.

[66]. Read, R. E. H. (1991), External fire spread: building separation and boundary distances, BR187, Building Research Establishment (BRE), U.K.

[67]. Johnson, C. and Tarlin, E. (1974), Incêndio (film), National Fire Protection Association and the National Bureau of Standards, U.S. Department of Commerce.

[68]. Routley, J., Jennings, C. and Chubb, M. (1991) High-rise office building fire, One Meridian Plaza, Philadelphia, Penn., February 23, 1991, United States Fire Administration, Technical Report Series, Report 049, FEMA, U.S.A.

[69]. Kenichi, I. and Sekizawa, A. (2005), Collapse mechanism of the Windsor building by fire in Madrid and the plan for its demolition process, International Workshop on Emergency Response and Rescue 2005, Taiwan.

[70]. Jacobs, A. (2009), Fire ravages renowned building in Beijing, The New York Times (Feb. 10th 2009), New York Edition, U.S.A.

[71]. Soares, G. (2010), Incêndio danifica hotel de 25 milhões, Diário de Notícias (10th March 2010), Portugal.

[72]. Cunha, S. (2010), Braga: Incêndio destrói parte da fachada de hotel, Correio da Manhã (9th March 2010), Portugal.

[73]. Cooke, G. (2009), Andraus high-rise, São Paulo, Brazil, February 1972 (presentation), Fire Safety Consultant London 2009, http://www.cookeonfire.com/pdfs/Andraus%20pdf.pdf [03/02/2010]

[74]. Tavares, R. M. (2008), Prescriptive codes vs. performance-based codes: Which one is the best fire safety code for the Brazilian context?, Safety Science Monitor 12 (1), Australia.

[75]. Beitel, J. and Iwankiw, N. (2002), Analysis of needs and existing capabilities for full-scale fire resistance testing, NIST GCR 02-843, National Institute of Standards and Technology, U.S.A.

[76]. Demers, D. P. (1981), Hotel fire, Las Vegas, NV, February 10, 1981, NFPA Fire Investigation, National Fire Protection Association, Quincy, MA, U.S.A.

[77]. O'Connor, D. J. (2008), Building facade or fire safety facade? CTBUH 8th World Congress, Dubai. Knight, K. (2009), Report to the Secretary of State by the Chief Fire and Rescue Adviser on the emerging issues arising from the fatal fire at Lakanal House, Camberwell on 3 July 2009, communities and Local Government, London, UK. ISBN 978-1-4098-1175-8.

[78]. Ingberg, S. H. (1928), Test of the severity of building fires, National Fire Protection Association Quarterly, 22 (1), pp.43-61.

[79]. Fujita, K. (1958), Characteristics of fire inside a non-combustible room and prevention of fire damage, Report 2 (2), Japanese Ministry of Construction, Building Research Institute, Tokyo, Japan.

[80]. Kawagoe, K. (1958), Fire behaviour in rooms, Report 27, Building Research Institute, Tokyo, Japan.

[81]. Thomas, P.H. and Law, M. (1972), The projection of flames from burning buildings, FRN 921, Fire Research Station, Borehamwood, U.K.

[82]. Thomas, P.H. and Heselden, A.J.M. (1972), Fully-developed fires in single compartments — A co-operative research programme of the Conseil International du Bâtiment, CIB Report No.20, FRN 923, Fire Research Station, Borehamwood, U.K.

[83]. Thomas, P.H. (1974), Fires in model rooms: CIB research programmes, Building Research Establishment Current Paper CP32/74, BRR, Borehamwood, U.K.

[84]. Magnusson, S.E. and Thelandersson, S. (1970), Temperature-time curves of complete process of fire development. Theoretical study of wood fuel fires in enclosed spaces, Acta Polytechnica Scandinavia, Civil Engineering and Building Construction Series 65, Stockholm, Sweden.

[85]. Babrauskas, V. and Williamson, R.B. (1978), Post-flashover compartment fires: Basis of a theoretical model, Fire and Materials 2 (2), pp.39-53.

[86]. Babrauskas, V. (1981), A closed-form approximation for post-flashover compartment fire temperatures, Fire Safety Journal 4, pp.63-73.

[87]. Yokoi, S. (1960), Study on the prevention of fire-spread caused by hot upward current, Building Research Institute, Report No. 34, Tokyo, Japan.

[88]. Webster, C.T. and Raftery, M.M. (1959), The burning of fires in rooms — Part II, FRN 401, Joint Fire Research Organization, Borehamwood, U.K.

[89]. Webster, C.T., Raftery, M.M., and Smith, P.G. (1961), The burning of fires in rooms — Part III, FRN 474, Joint Fire Research Organization, Borehamwood, U.K.

[90]. Webster, C.T. and Smith, P.G. (1964), The burning of well-ventilated compartment fires — Part IV brick compartment, 2.4 m (8 ft) cube, FRN 578, Joint Fire Research Organization, Borehamwood, U.K.

[91]. Thomas, P. H. (1961), On the heights of buoyant flames, FRN 489, Fire Research Station, Borehamwood, U.K.

[92]. Seigel, L.G. (1969), The projection of flames from burning buildings, Fire Technology 5 (1), pp.43-51.

[93]. Heselden, A.J.M., Smith, P.G. and Theobald, C.R. (1966), Fires in a large compartment containing structural steelwork — Detailed measurements of fire behaviour, FRN 646, Fire Research Station, Borehamwood, U.K.

[94]. Gross, D. (1967), Field burnout tests of apartment dwelling units, Building Science Series 10, National Bureau of Standards, Sept. 29.

[95]. Underwriters Laboratories, Inc. (1970), Fire test of Walt Disney World unitized guest room, report for United States Steel Corporation, Underwriters Laboratories, Inc., July 1, Illinois, U.S.A.

[96]. Butcher, E.G., Chitty, T.B. and Ashton, L.A. (1966), The temperature attained by steel in building fires, Fire Research Technical Paper 15, HMSO, London, U.K.

[97]. Underwriters Laboratories, Inc. (1975), Fire severity at the exterior of a burning building, American Iron and Steel Institute, April 3, Washington, D.C., U.S.A.

[98]. Stromdahl, I. (1972), The Tranas Fire Tests — Field studies of heat radiation from fires in a timber structure, National Swedish Institute for Building Research, Document D3, Stockholm, Sweden.

[99]. Kordina, K. (1978), Brandversuche Lehrte — Schriftenreihe "Bau und Wohnforschung" des Bundesministers für Raumordnung, Bauwesen und Städtebau, Bonn, Germany.

[100]. Buchanan, A.H. (2002), Structural Design for Fire Safety, John Wiley & Sons, U.K. ISBN: 0-471-89060-X.

[101]. Thomas, P.H. (1963), The size of flames from natural fires, Symposium (International) on Combustion 9 (1), pp.844-859.

[102]. Bullen, M.L. and Thomas, P.H. (1979), Compartment fires with non-cellulosic fuels, Proceedings of the 17th Symposium (International) on Combustion 17 (1), pp.1139-1148, The Combustion Institute, U.S.A.

[103]. Bohm, B. and Rasmussen, B.M. (1987), The development of a small-scale fire compartment in order to determine thermal exposure inside and outside buildings, Fire Safety Journal 12 (2), pp.103-108.

[104]. Oleszkiewicz, I. (1989), Heat transfer from a window fire plume to a building facade, HTD 123, ASME Collected papers on Heat Transfer, Book No. H00526, pp.163-170.

[105]. Oleszkiewicz, I. (1990), Fire exposure to exterior walls and flame spread on combustible cladding, Fire Technology 26, pp.357-375.

[106]. Oleszkiewicz, I. (1991), Vertical separation of windows using spandrel walls and horizontal projections, Fire Technology 27, pp.334-340.

[107]. Gottuk, D.T., Roby, R.J. and Beyler, C.L. (1992), A study of carbon monoxide and smoke yields from compartment fires with external burning, Symposium (International) on Combustion 24 (1), pp.1729-1735.

[108]. Ohmiya, Y., Yusa, S., Suzuki, J.I., Koshikawa, K. and Delichatsios, M.A. (2003), Aerodynamics of fully involved enclosure fires having external flames, Proceedings of the Fourth International Seminar on Fire and Explosion Hazards, University of Ulster, Belfast, U.K.

[109]. Ohmiya, Y., Tanaka, T. and Wakamatsu, T. (1998), A room fire model for predicting fire spread by external flames, Fire Science and Technology 18 (1), pp.11-21.

[110]. Ohmiya, Y., Hori, Y., Sagimori, K. and Wakamatsu, T. (2000), Predictive method for properties of flame ejected from an opening incorporating excess fuel, Proceedings of the Fourth Asia-Oceania Symposium on Fire Science and Technology, pp. 375-386.

[111]. Lee, Y.-P., Delichatsios, M.A. and Silcock, G.W.H. (2007), Heat fluxes and flame heights in facades from fires in enclosures of varying geometry, Proceedings of the Combustion Institute 31 (2), pp.2521-2528.

[112]. Lee, Y.-P. (2006), Heat fluxes and flame heights in external facades from enclosure fires, Ph.D. Thesis, The University of Ulster, Belfast, U.K.

[113]. Goble, K. (2007), Height of flames projecting from compartment openings, Masters Thesis, The University of Canterbury, Christchurch, New Zealand.

[114]. Klopovic, S. and Turan, Ö.F. (1998), Flames venting externally during full-scale flashover fires: Two sample ventilation cases, Fire Safety Journal 31 (2), pp.117-142.

[115]. Klopovic, S. and Turan, Ö.F. (2001) A comprehensive study of externally venting flames— Part I: External plume characteristics for through-draught and no-through-draught ventilation conditions and repeatability, Fire Safety Journal 36 (2), pp.99-133.

[116]. Klopovic, S. and Turan, Ö.F. (2001), A comprehensive study of externally venting flames— Part II: Plume envelope and centre-line temperature comparisons, secondary fires, wind effects and smoke management system, Fire Safety Journal 36 (2), pp.135-172.

[117]. Suzuki, T., Sekizawa, A., Satoh, H., Yamada, T., Yanai, E., Kurioka, H. and Kimura, Y. (1999), An experimental study of ejected flames of a high-rise building Part I, National Research Institute of Fire and Disaster No.88, Japan, pp.51-63.

[118]. Takahashi, K., Suzuki, T., Yamada, T., Yanai, E., Sekizawa, A., Kurioka, H. and Satoh, H. (2001), Flame configuration derived from video camera picture—An experimental study of ejected flames of a high-rise building Part II, National Research Institute of Fire and Disaster No.91, Japan, pp.48-58.

[119]. Yamada, T., Takahashi, K., Yanai, E., Suziki, T., Sekizawa, A., Satoh, H. and Kurioka, H. (2001), Heat flux to surface walls above fire floor level—An experimental study of ejected flames of a high-rise building Part III, National Research Institute of Fire and Disaster No.88, Japan, pp.59-68.

[120]. Hakkarainen, T. and Oksanen, T. (2002), Fire safety assessment of wooden facades, Fire and Materials 26, pp.7-27.

[121]. Sugawa, O., Momita, D. and Takahashi, W. (1996), Flow behaviour of ejected fire flame/plume from an opening effected by external side wind, Proceedings of the 5th International Symposium on Fire Safety Science, pp.249-260.

[122]. ARUP, Innovator's hall of fame, Short biography of "Margaret Law", http://info.arup.com/arup/feature.cfm?pageid=8984 [accessed 05/03/09]

[123]. Incropera, F.P. and DeWitt, D.P. (2002), Fundamentals of Mass and Heat Transfer, Fifth Edition, John Wiley & Sons, Inc., U.S.A. ISBN: 0-471-38650-2.

[124]. Arnault, P., Ehm, H. and Kruppa, J. (1973), Rapport experimental sur les essays avec des feux naturels executes dans la petite installation, CECM 3/73-11-F, CTICM, Puteaux, France.

[125]. Howell, J.R. (1982), A catalog of radiation heat transfer configuration factors, 3rd Edition, The University of Texas at Austin, Austin, Texas, U.S.A. http://www.engr.uky.edu/rtl/Catalog/ [accessed 07/09/2008]

[126]. McGuire, J.H. (1952), The calculation of heat transfer by radiation, FRN 20, Fire Research Station, Borehamwood, U.K.

[127]. BSI (1972), BS476: Part 20: 1987—Fire tests on building materials and structures: Methods for determination of the fire resistance of elements of construction (general principles), British Standards, U.K.

[128]. BSI (1972), BS476: Part 21: 1987—Fire tests on building materials and structures: Methods for determination of the fire resistance of load bearing elements of construction, British Standards, U.K.

[129]. BSI (1972), BS476: Part 8: 1972—Fire tests on building materials and structures: Test methods and criteria for the fire resistance of elements of building construction, British Standards, U.K.

[130]. Seigel, L.G. (1970), Fire test of an exposed steel spandrel girder, Materials Research and Standards, MTRSA 10 (2), pp.10-13.

[131]. Arnault, P., Ehm, H. and Kruppa, J. (1974), Evolution des temperatures dans des poteaux exterieurs soumis a des incendices, CECM 3-74/7F, CTICM, Puteaux, France.

[132]. Law, M. (1981/2), Notes on the external fire exposure measured at Lehrte, Fire Safety Journal 4 (4), pp.243-246, Elsevier Science, U.K.

[133]. Ashton, L.A. and Malhotra, H.L. (1960), External walls on buildings, I—The protection of openings against spread of fire from storey to storey, FRN 836, Borehamwood, U.K.

[134]. Moulen, A.W. (1971), Horizontal projections in the prevention of spread of fire from storey to storey, Report TR52/75/397, Commonwealth Experimental Building Station, Australia.

[135]. Harmathy, T.Z. (1979), Design to cope with fully developed compartment fires, in Design of Buildings for Fire Safety, (Eds.) E.E. Smith and T.Z. Harmathy, American Society for Testing and Materials, STP 685, pp.198-276.

[136]. Becker, R. (2002), Structural behaviour of simple steel structures with non-uniform longitudinal temperature distributions under fire conditions, Fire Safety Journal 37 (5), pp.495-515.

[137]. Gillie, M., Usmani, A.S. and Rotter, J.M. (2002), A structural analysis of the Cardington British Steel Corner Test, Journal of Construction Steel Research 58 (4), pp.427-442.

[138]. Welch, S. and Lennon, T. (2001), Comments on Eurocode 1—Actions on structures, Part 1- 2: General actions—actions on structures exposed to fire, Amended FINAL DRAFT (Stage34), 24 August 2001, BRE, U.K.

[139]. Reszka, P., Abecassis Empis, C., Biteau, H., Cowlard, A., Steinhaus, T., Fletcher, I.A., Fuentes, A., Gillie, M. and Welch, S. (2007), Experimental layout and building description, Chapter 2, The Dalmarnock Fire Tests: Experiments and Modelling, (Eds.) Rein, G., Abecassis Empis, C. and Carvel, R., School of Engineering and Electronics, The University of Edinburgh, U.K. ISBN 978-0-9557497-0-

[140]. Cowlard, A., Steinhaus, T., Abecassis Empis, C. and Torero, J.L. (2007), Test Two: The 'Controlled fire', Chapter 4, The Dalmarnock Fire Tests: Experiments and Modelling, (Eds.) Rein, G., Abecassis Empis, C. and Carvel, R., School of Engineering and Electronics, The University of Edinburgh, U.K. ISBN 978-0-9557497-0-4.

[141]. Amundarain, A.A., Torero, J.L., Usmani, A., Al-Remal, A.M. (2007), Assessment of the thermal efficiency, structure and fire resistance of lightweight building systems for opti-mised design, PhD Thesis, School of Engineering, The University of Edinburgh, U.K. http://hdl.handle.net/1842/2128 [accessed 01/04/2008].

[142]. Jowsey, A., Torero, J.L., and Lane, B. (2007), Heat transfer to the structure during the Fire, Chapter 7, The Dalmarnock Fire Tests: Experiments and Modelling, (Eds.) Rein, G., Abecassis Empis, C. and Carvel, R., School of Engineering and Electronics, The University of Edinburgh, U.K. ISBN 978-0-9557497-0-4.

[143]. Gillie, M. and Stratford, T. (2007), Behaviour of the structure during the fire, Chapter 8, The Dalmarnock Fire Tests: Experiments and Modelling, (Eds.) Rein, G., Abecassis Empis, C. and Carvel, R., School of Engineering and Electronics, The University of Edinburgh, U.K. ISBN 978-0-9557497-0

[144]. Stratford, T., Gillie, M. and Chen, J-F. (2007), The performance of fibre reinforced polymer strengthening in the fire, Chapter 9, The Dalmarnock Fire Tests: Experiments and Modelling, (Eds.) Rein, G., Abecassis Empis, C. and Carvel, R., School of Engineering and Electronics, The University of Edinburgh, U.K. ISBN 978-0-9557497-0-4.

[145]. Welch, S., Jowsey, A., Deeny, S., Morgan, R. and Torero, J.L. (2007), BRE large compart-ment fire tests—characterising post-flashover fires for model validation, Fire Safety Journal 42 (8), pp.548-567. doi:10.1016/j.firesaf.2007.04.002.

[146]. Jin, T. (2008), Visibility and human behaviour in fire smoke, The SFPE Handbook of Fire Protection Engineering, 4th Edition, (Eds.) P.J. DiNenno et al., National Fire Protection Association, Massachusetts, U.S.A., pp. 2-54—2-66.

[147]. Abecassis Empis, C., Cowlard, A., Welch, S. and Torero, J.L. (2007), Test One: The 'Uncontrolled' fire, Chapter 3, The Dalmarnock Fire Tests: Experiments and Modelling, (Eds.) Rein, G., Abecassis Empis, C. and Carvel, R., School of Engineering and Electronics, The University of Edinburgh, U.K. ISBN 978-0-9557497-0-4.

[148]. Mulholland, G.W. (2002), Smoke production and properties, The SFPE Handbook of Fire Protection Engineering, 3rd Edition, (Eds.) P.J. DiNenno et al., National Fire Protection Association, Massachusetts, U.S.A., pp. 2-291—2-302.

[149]. Ingason, H. and Wickstrom, U. (2007), Measuring incident radiant heat flux using the plate thermometer, Fire Safety Journal 42 (2), pp.161-166.

[150]. ASTM (2005), ASTM E459-05: Standard test method for measuring heat transfer rate using a thin-skin calorimeter, ASTM International, U.S.A. DOI: 10.1520/E0459-05.

[151]. McCaffrey, B.J., Heskestad, G. (1979), A robust bidirectional low-velocity probe for flame and fire application, Combustion and Flame 26 (1), pp.125-127.

[152]. Cowlard, A. (2009), Sensor and model integration for the rapid prediction of concurrent flow flame spread, PhD Thesis, School of Engineering, The University of Edinburgh, U.K. http://hdl.handle.net/1842/2753 [accessed 18/09/2009].

[153]. Huggett, C. (1980), Estimation of rate of heat release by means of oxygen consumption measurements, Fire and Materials 4, pp.61-65.

[154]. Kawagoe, K. and Sekine, T. (1963), Estimation of fire temperature-time curve in rooms, BRI Occasional Report 11, Building Research Institute, Tokyo, Japan.

[155]. BSI (1993), BS 476: Part 33: 1993 (ISO 9705:1993)—Fire tests on building materials and structures: full-scale room test for surface products, British Standards, U.K.

[156]. Babrauskas, V. (1981), Will the second item ignite?, National Bureau of Standards, Gaithersburg, MD, U.S.A. NBSIR 81-2271.

[157]. Joshi, A. and Pagni, P.J. (1990), Thermal analysis of effect of a compartment fire on win-dow glass, Fire Research and Safety, 11th Joint Panel Meeting, (Eds.) Jason, N. H., Cramer, D. M., Berkeley, CA, pp. 233-252.

[158]. Mowrer, F.W. (1997), Window breakage induced by exterior fires, Proceedings of the 2nd International Conference on Fire Research and Engineering (ICFRE2), NIST and SFPE sponsored, Gaithersburg, MD, U.S.A.

[159]. Desanghere, S. (2007), Development of a simplified model aimed at predicting external members heating conditions, Interflam 2007—Proceedings of the Eleventh International Conference (2), Interscience Communications, London, UK., pp. 955-966.

# References

1. Wade, C.A., Clampett, J.C. (2000) "Fire Performance of Exterior Claddings". Sydney, Australia: Fire Code Reform Centre, April 2000. Report No.: Project Report FCRC PR 00-03.
2. Hu, L.H., Lu, K.H., Tang, F., Delichatsios, M., He, L.H., editors. (2013) "Effects of side walls on facade flame entrainment and flame height from opening in compartment fires". 1st International Seminar for Fire Safety of Facades; 2013 14-15 November 2013; Paris, France: MATEC Web Conferences. Available from: http://www.matec-conferences.org/index. php?option=com_toc&url=/articles/matecconf/abs/2013/07/contents/contents.html.
3. (2000) "Guide to Exterior Insulation and Finish System Construction". Falls Church, VA, USA.
4. "PE Aluminium composite panel image from manufacturer website" [cited 2014 17 March 2014]. Available from: http://www.alumcompositepanel.com/pe-aluminum-panel.html.
5. "HPL construction image from Alumina Elit website": Alumina Elit; [cited 2014 17 March 2014]. Available from: http://www.aluminaelit.com/index.pl?todo=products&lang=en.
6. Griffin, G.J., Bicknell, A.D., Bradbury, G.P., White, N. (2006) "Effect of Construction Method on the Fire Behaviour of Sandwich Panels with Expanded Polystyrene Cores in Room Fire Tests". Journal of Fire Sciences. 2006;24(July 2006):275-94.
7. Cooke, G.M.E. (2000) "Sandwich panels for external cladding — fire safety issues and implications for the risk assessment process". Cowbridge, UK: Eurisol — UK Mineral Wool Association Report. Available from: http://www.cookeonfire.com/pdfs/eurisolgreenreport.pdf.
8. (2013) "Linear Facade Systems 2013 Rain Screen Catalogue". UK: Euroclad Facades Limited. Available from: http://www.euroclad-architectural.co.uk/media/10477/euroclad_facades_linear_rainscreen.pdf.
9. England, P., Eyre, M. (2011) "Fire Safety Engineering Design of Combustible Façades". Melbourne, Australia: Exova Warringtonfire Aus Pty Ltd, Report No.: Project No: PNA217-1011. Available from: http://www.fwpa.com.au/sites/default/files/Fire%20 Safety%20Engineering%20Design%20of%20Combustible%20Facades%20-%20Issue%201. pdf.
10. Standards Australia. (2005) "AS 1530.4-2005 : Methods for fire tests on building materials, components and structures — Fire-resistance test of elements of construction". Sydney, Australia: SAI Global; 2005.
11. ISO. (2002) "ISO 13785-1:2002 Reaction-to-fire tests for façades — Part 1: Intermediate-scale test". Geneva, Switzerland: International Organization for Standardization; 2002.
12. ISO. (2002) "ISO 13785-2:2002 Reaction-to-fire tests for façades — Part 2: Large-scale test". Geneva, Switzerland: International Organization for Standardization; 2002.

13. Thomas, P.H., Law, M. (1972) "The projection of flames from burning buildings". Borehamwood, U.K.: Fire Reserach Station, Document No.: FRN 921.

14. Law, M. (1978) "Fire safety of external building elements—The design approach,". Engineering Journal. 1978;Second Quarter:59-74.

15. Langdon-Thomas, G.J., Law, M. (1966) "Fire and the external wall". Joint Fire Research O, editor. London,: H.M.S.O.; 1966. [5], 16 p.

16. Meacham, B., Poole, B., Echeverria, J., Cheng, R. (2012) "Fire Safety Challenges of Green Buildings". Quincy, MA, USA: Worcester Polytechnic Institute, November 2012.

17. Colwell, S., Baker, T. (2013) "Fire performance of external thermal insulation for walls of multistorey buildings". Garston, Watford, UK: IHS BRE Press, Report No.: 978-1-84806-234-4 Document No.: BR 135.

18. Mikkola, E., Hakkarainen, T., Matala, A. (2013) "Fire Safety of EPS ETICS in residential multi-storey buildings". Finland: VTT, 26/6/2013. Document No.: Research Report VTT-R-04632-13.

19. MATEC Web Conferences, editor (2013) "Proceedings 1st International Seminar for Fire Safety of Facades" 2013 14-15 November 2013; Paris, France: MATEC Web Conferences. Available from: http://www.matec-conferences.org/index.php?option=com_toc&url=/articles/matecconf/abs/2013/07/contents/contents.html.

20. Alpert, R.L., Davis, R.J. (2002) "Evaluation of Exterior Insulation and Finish System Fire Hazard for Commercial Applications". Journal of Fire Protection Engineering. 2002 November 1, 2002;12(4):245-58.

21. Delichatsios, M. (2014) "Enclosure and Facade Fires: Physics and Applications—Plenary paper". IAFSS Conference: In Press. p. February 2014.

22. EMPIS, C.A. (2010) "Analysis of the compartment fire parameters influencing the heat flux incident on the structural façade". Available from: http://www.era.lib.ed.ac.uk/bitstream/1842/4188/1/AbecassisEmpis_HeatFluxFacade_PhD2010.pdf.

23. Oleszkiewicz, I. (1989) "Heat transfer from a window fire plume to a building facade". ASME Collected papers on Heat Transfer,. Book No. H00526. p. 163-70.

24. Oleszkiewicz, I. (1990) "Fire exposure to exterior walls and flame spread on combustible cladding". Fire Technology. 1990 1990/11/01;26(4):357-75. English.

25. Ryan, J. (2011) "Vertical Separation Distance in Multi-Storey Buildings, M.Sc. Fire Safety Engineering". Jordanstown, NI, UK: FireSERT, Faculty of Built Environment, University of Ulster.

26. Zhang, J., Delichatsios, M., McKee, M., Ukleja, S., Pagella, C. (2011) "Experimental study of burning behaviors of intumescent coatings and nanoparticles applied on flaxboard". Journal of Fire Sciences. 2011;29(519).

27. Hall Jr., J.R., Harwood, B. (1989) "The national estimates approach to U.S. fire statistics". Fire Technology. 1989 May 1989;Volume 25(Issue 2):pp 99-113.

28. (2010) "National Fire Incident Reporting System—complete reference guide". July 2010.

29. ICC. (2012) "2012 International Building Code". USA.

30. NFPA. (2012) "NFPA 5000, Building Construction and Safety Code". Quincy, MA

31. McGrattan, K.B., Hostikka, S., Floyd, J., Baum, H., Rehm, R., Mell, W., et al. (2008) "Fire Dynamics Simulator (Version 5) Technical Reference Guide Volume 1 Mathematical Model". Washington: National Institute of Standards and Technology, 2008. Report No.: NIST Special Publication 1018-5.

32. McGrattan, K.B., Klein, B., Hostikka, S., Floyd, J. (2008) "Fire Dynamics Simulator (Version 5) Users Guide". Washington: National Institute of Standards and Technology, 2008. Report No.: NIST Special Publication 1019-5.

33. McGrattan, K.B., McDerrmott, R., Hostikka, S., Floyd, J. (2010) "Fire Dynamics Simulator (Version 5) Technical Reference Guide Volume 3 Validation". Washington: National Institute of Standards and Technology, 2010. Report No.: NIST Special Publication 1018-5.

34. (2003-2007) "NSW Fire Brigades Annual Statistical Reports 2007-2007". Sydney, NSW, Australia. Available from: http://www.fire.nsw.gov.au/page.php?id=171.

35. (2005-2010) "The New Zealand Fire Service Emergency Incident Statistics 2005-2010". Wellington, New Zealand. Available from: http://www.fire.org.nz/About-Us/Facts-and-Figures/Documents/Stats-09-10s.pdf.

36. HAJPÁL, D.M., editor (2012) "Analysis of a tragic fire case in panel building of Miskolc". Integrated Fire Engineering and Response; 2012; Malta. Available from: http://lacoltulstrazii.files.wordpress.com/2012/10/analysis-of-a-tragic-fire-case-in-panel-building-miskolc-hungary.pdf.

37. Post, N.M., Illia, T. (2008). The Construction Weekly—Engineering News Record. 2008;February /March 2008(Decorative material under investigation):pgs 12-4.

38. Beitel, J.J., Douglas H. Evans. (2011) "The Monte Carlo Exterior Facade Fire". SPFE Fire Protection Engineering. 2011 01/10/2011.

39. Duval, R. (2008) "Monte Carlo Hotel Fire". NFPA Journal. 2008 May/Jone 2008.

40. Oleszkiewicz, I. (1990) "Fire performance of external insulation system : observations made after the fire at 393 Kennedy Street, Winnipeg, Manitoba, January 10, 1990".

41. Oleszkiewicz, I. (1995) "Fire testing and real fire experience with EIFS in Canada". ASTM Special Technical Publication. 1995 (1187):129-39.

42. (2010) "Seven die in fire in immigrant hostel in Dijon, France". BBC News Europe. 2010 14 November 2010 Available from: http://www.bbc.co.uk/news/world-europe-11752303.

43. Broemme, A. (2005) "Berlin: Verheerender Fassadenbrand". Deutsche Feuerwehr-Zeitung Brandschutz. 2005 August, 2005.;number 6.

44. Bong, F.N.P. (2000) "Fire Spread on Exterior Walls—Fire Engineering Masters Thesis". New Zealand: University of Canterbury, Report No.: Fire Engineering Report Number 2000/1. Available from: http://www.civil.canterbury.ac.nz/fire/pdfreports/Felix1.pdf.

45. Messerschmidt, B. (2012) "RE: Another Exterior Wall Fire.". Received by: White N. Email containing power point slides of Rockwool investigation of facade fire in Roubaix, France 14th May 2012. Received: Thu 11/07/2013 11:27 PM.

46. (2013) "l'incendie tour mermoz pompiers de Roubaix" [Movie]. YouTube; 2013 [cited 2013 19 July 2013]. Footage of Mermoz Tower Fire, Roubaix, France]. Available from: http://www.youtube.com/watch?v=j4mIBQnUAfQ.

47. (2012) "Spectacular High-Rise Fire in France" 2012 [cited 2013 19 July 2013]. Blog report on Mermoz Tower fire, Roubaix, France]. Available from: http://firegeezer.com/2012/05/15/spectacular-high-rise-fire-in-france/.

48. (2012) "High-rise blaze in 18-storey block in Roubaix, France" 2012 [cited 2013 19 July 2013]. Blog report on Mermoz Tower fire, Roubaix]. Available from: http://www.blog.plumis.co.uk/2012/05/high-rise-blaze-in-18-storey-block-in.html.

49. Baldwin, D., Leon, J.P.d. (2012) "Tower cladding in UAE fuels fire". Gulf News. 2012 May 2, 2012. Available from: http://gulfnews.com/news/gulf/uae/housing-property/tower-cladding-in-uae-fuels-fire-1.1016836.

50. Abdullah, A. (2012) "Cigarette butt caused blaze at Al Tayer Tower". Khaleej Times. 2012 6 June 2012. Available from: http://khaleejtimes.com/kt-article-display-1.asp?xfile=/data/nationgeneral/2012/June/nationgeneral_June49.xml&section=nationgeneral.

51. (2012) "Two Serious Fire Outbreaks in Dubai Towers". FIRE Middle East. 2012. Available from: http://www.firemiddleeastmagazine.com/news/article/61.

52. Leon, J.P.d., Barakat, N. (2012) "Fire in Tecom building leaves seven families homeless". Gulf News. 2012 October 6, 2012. Available from: http://gulfnews.com/news/gulf/uae/emergencies/fire-in-tecom-building-leaves-seven-families-homeless-1.1085854.

53. Croucher, M. (2012) "Residents of Dubai's Tamweel Tower relive fire ordeal". TheNationalUAE. 2012 Nov 19, 2012 Available from: http://www.thenational.ae/news/uae-news/residents-of-dubais-tamweel-tower-relive-fire-ordeal#ixzz2evw6daTm.

54. Croucher, M. (2012) "Aggressive changes to UAE fire-safety code after hundreds left homeless". TheNationalUAE. 2012 Nov 26, 2012 Available from: http://www.thenational.ae/news/uae-news/aggressive-changes-to-uae-fire-safety-code-after-hundreds-left-homeless#ixzz2evws04Jw.

55. (2012) "Fire breaks out at Tamweel Tower in Jumeirah Lake Towers—Police begin probe into cause of fire that displaced hundreds of people". Gulfnewscom. 2012. Available from: http://gulfnews.com/news/gulf/uae/emergencies/fire-breaks-out-at-tamweel-tower-in-jumeirah-lake-towers-1.1106387.

56. "Haeundae Highrise on Fire—Busan Marine City Burns" [cited 2013 8/10/2013]. Available from: http://koreabridge.net/post/haeundae-highrise-fire-busan-marine-city-burns.

57. Mee-yoo, K. (2010) "High-rise apartments defenseless against fire". The Korea Times. 2010 03/10/2010. Available from: http://www.koreatimes.co.kr/www/news/nation/2010/10/113_73908.html.

58. Young-sun Kim, M.M., Yoshifumi Ohmiya. (2011) "Fire Examination of Superhigh-Rise Apartment Building "Wooshin Golden Suites" in Busan, Korea". Fire Science and Technology. 2011;Vol 30(No 3):81-90.

59. Kim, Y.-s., Mizuno, M., Ohmiya, Y. (2013) "FIRE INCIDENT REPORT (2) "Wooshin Golden suites"in Busan" 2013 [cited 2013 30/10/2013]. Available from: http://www.tus-fire.com/?p=1761.

60. FOLEY, J.M. (2010) "Modern Building Materials Are Factors in Atlantic City Fires". Fire Engineering. 2010 05/01/2010.

61. Taber, B.C., Gibbs, E. (2007) "Full-Scale exterior wall fire test on a Composites Gurea exterior wall panel system". Canada: National Research Council of Canada, October 25, 2007. Document No.: Report No: B-4198.1.

62. (2013) "2010 Shanghai fire" Wikipedia2013 [cited 2013 22 July 2013]. Available from: http://en.wikipedia.org/wiki/2010_Shanghai_fire#cite_note-8.

63. (2002) "TIP TOP BAKERY FIRE, FAIRFIELD—Post Incident Summary Report". New South Wales Fire Brigades, August 2002. Document No.: PIA NO 011/02.

64. Harwood, J., Hume, B. (1997) "Fire Safety of Sandwich Panels Summary Report". UK: Central Fire Brigades Advisory Council, Joint Committee on Fire Research, Document No.: Research Report Number 76.

65. (1999) "Memorandum by the Fire Safety Development Group (ROF 26)": Parliament UK Publications and Records; 1999 [cited 2013 1/11/2013]. Available from: http://www.publications.parliament.uk/pa/cm199899/cmselect/cmenvtra/741/9072003.htm.

66. GRIFFITH, M. (1997) "Spectacular fire engulfs casino's plastic facade". The Associated Press. 1997 Oct. 1, 1997 Available from: http://www.apnewsarchive.com/1997/Spectacular-fire-engulfs-casino-s-plastic-facade/id-dd2ef61811fd3ff74766d088415daaf6.

67. Evans, D.H. (2008) "Presentation tiitled "The Monte Carlo Facade Fire" dated January 25, 2008". Received by: White N.

68. (1998) "Four alarm fire dmages Palace Station Hotel". Las Vegas Firenews, Las Vegas Fire Department. 1998 October 1998;Volume 7(No 4).

69. (2013) "Grozny skyscraper catches fire". New York Daily News. 2013. Available from: http://www.nydailynews.com/news/world/tallest-building-grozny-chechnya-engulfed-fire-article-1.1306655.

70. (2013) "Fire in Grozny City tower put out, no casualties reported". ITAR-TASS News Agency. 2013. Available from: http://en.itar-tass.com/russia/691795.

71. (2013) "Fire engulfs high-rise building in Russia—GROZNY World News" 2013. Available from: http://www.youtube.com/watch?v=9-fN8KKmBbI.

72. (2013) "Beijing Television Cultural Center fire": Wikipedia; 2013 [cited 2013 20 July 2013]. Available from: http://en.wikipedia.org/wiki/Beijing_Television_Cultural_Center_fire.

73. Kobayashi, K. (2014) "The Effects of Revisions to the Fire Regulations on Building Fire Damages". Journal of Fire Science and Technology. 2014;Center for Fire Science and Technology, Research Institute for Science and Technology, Tokyo University of Science.

74. Hokugo, A., Hasemi, Y., Hayashi, Y., Yoshida, M. (2000) "Mechanism for the Upward Fire Spread through Balconies Based on an Investigation and Experiments for a Multi-story Fire in High-rise apartment Building". Fire Safety Science. 2000;67:649-60.

75. Australian Building Codes Board. (2013) "National Construction Code Series Volume 1, Building Code of Australia 2013, Class 2 to 9 Buildings". Canberra: Australian Building Codes Board; 2013.

76. Australian Building Codes Board. (2013) "National Construction Code Series Volume 2, Building Code of Australia 2013, Class 1 and 10 Buildings". Canberra: Australian Building Codes Board; 2013.

77. Ministry of Business, Innovation & Employment. (2014) "Acceptable Solutions and Verification Methods". New Zealand. Available from: http://www.dbh.govt.nz/compliance-documents#C.

78. HM Government. (2013) "The Building Regulations 2010, Fire Safety Approved Document B". UK. Available from: http://www.planningportal.gov.uk/buildingregulations/approveddocuments/partb/bcapproveddocumentsb/.

79. BSI. (2002) "BS 8414-1:2002 Fire performance of external cladding systems. Test methods for non-loadbearing external cladding systems applied to the face of a building". UK.: British Standards Institute; 2002.

80. BSI. (2005) "BS 8414-2:2005 Fire performance of external cladding systems Test method for non-loadbearing external cladding systems fixed to and supported by a structural steel frame". UK.: British Standards Institute; 2005.

81. BSI. (1989) "BS 476-6:1989+A1:2009 Fire tests on building materials and structures Method of test for fire propagation for products". UK.: British Standards Institute; 1989.

82. BSI. (1982) "BS 476-11:1982 Fire tests on building materials and structures Method for assessing the heat emission from building materials". UK.: British Standards Institute; 1982.

83. CEN. (2007) "EN13501-1:2007: Fire classification of construction products and building elements-Part1: Classification using data from reaction to fire tests". European Committee for Standardization; 2007.

84. LPCB. (2006) "LPS 1181: PART 4: ISSUE 1, Series of Fire Growth Tests for LPCB Approval and Listing of Construction Product Systems, Part Four: Requirements and Tests for External Thermal Insulated Cladding Systems with rendered finishes (ETICS) or Rain Screen Cladding systems (RSC) applied to the face of a building". UK.: Loss Prevention Cerification Board.

85. Strömgren, M., Albrektsson, J., Johansson, A., Almgren, E. (2013) "Comparative analysis of façade regulations in the Nordic countries". 1st International Seminar for Fire Safety of Facades; Paris, France: MATEC Web of Conferences Available from: http://dx.doi.org/10.1051/matecconf/20130901003.

86. ICC. (2012) "2012 International Building Code (IBC)". USA: International Code Council.

87. NFPA. (2012) "NFPA 5000: Building Construction and Safety Code®". Quincy, MA, USA: National Fire Protection Association.

88. NFPA. (2012) "NFPA 285: Standard Fire Test Method for Evaluation of Fire Propagation Characteristics of Exterior Non-Load-Bearing Wall Assemblies Containing Combustible Components". Quincy, MA, USA: National Fire Protection Association.

89. ASTM. (2013) "ASTM E84-13a: Standard Test Method for Surface Burning Characteristics of Building Materials". West Conshohocken, PA, United States: ASTM International; 2013.

90. UL. (2008) "UL 723: Standard for Test for Surface Burning Characteristics of Building Materials". USA: Underwriters Laboratories; 2008.

91. ASTM. (2013) "ASTM D1929-13a: Standard Test Method for Determining Ignition Temperature of Plastics". West Conshohocken, PA, United States: ASTM International; 2013.

92. FM Approvals. (2007) "ANSI FM 4880-2001(R2007): American National Standard for Evaluating; A) Insulated wall or wall & roof/ceiling assemblies; B) Plastic interior finish materials; C) Plastic exterior building panels; D) Wall/Ceiling coating systems; E) Interior or Exterior Finish Systems". Norwood, MA, USA: American National Standards Institute and FM Approvals; 2007.

93. NRC. (2010) "National Fire Code of Canada (NFC) 2010". Ottawa, Ontario, Canada: National Research Council Canada.

94. Standards Council of Canada. (2013) "CAN/ULC-S134-13: Standard Method of Fire Test of Exterior Wall Assemblies". Ottawa, Ontario, Canada: Standards Council of Canada and Underwriters Laboratories Canada; 2013.

95. (2011) "UAE Fire and Life Safety Code of Practice". United Arab Emirates: General Headquarters of Civil Defence, United Arab Emirates.

96. (2013) "UAE Fire and Life Safety Code of Practice: ANNEXURE. A.1.21. FIRE STOPPING, EXTERIOR WALL CLADDING AND ROOFING SYSTEMS". United Arab Emirates: General Headquarters of Civil Defence, United Arab Emirates. Available from: http://www.dcd.gov.ae/download/preventionsafety/ANNEXURE_A.1.21.%20PASSIVE%20FIRE%20STOPPING_EXTERIOR%20WALL%20CLADDING%20AND%20ROOFING%20SYSTEMS.pdf.

97. SCDF. (2013) "Singapore Fire Code 2013". Singapore: Singapore Civil Defence Force. Available from: http://www.scdf.gov.sg/content/scdf_internet/en/building-professionals/publications_and_circulars/fire-code-2013.html.

98. Legal Research Board. (2007) "Uniform Building By-Laws 1984 (UBBL) (as at 20th November 2007)". Malaysia: 2007.

99. (2006) "Guide to Fire Protection in Malasia". Dato' Hamzah Bin Abu Bakar ed. Kuala Lumpur, Malaysia: The Institution of Fire Engineers (UK) Malaysia Branch (IFEM).

100. ASTM. (2009) "E2568 – 09: Standard Specification for PB Exterior Insulation and Finish Systems". West Conshohocken, PA, United States: ASTM International; 2009.

101. ANSI. (2001) "ANSI/EIMA 99-A-2001: American National Standard for Exterior Insulation and Finish Systems (EIFS)". New York, USA: American National Standards Institute and EIMA-EIFS Industry Members Association; 2001.

102. EOTA. (2011) "ETAG 004: GUIDELINE FOR EUROPEAN TECHNICAL APPROVAL of EXTERNAL THERMAL INSULATION COMPOSITE SYSTEMS WITH RENDERING". Brussels: European Organisation for Technical Approvals; 2011.

103. Macdonald, N.J. (2012) "A comparison of BS 8414-1 & -2, draft DIN 4102-20, ISO 13785-1 & -2, EN 13823 and EN ISO 11925-2". Herts, UK: BRE Global, 28 June 2012. Document No.: Report number CC 275194 issue 2.

104. Hansbro, J. (2010) "NFPA 285-2006 Approval for wall assemblies using foam plastic insulation". Interface. 2010 January 2010:34-6.

105. SP. (1994) "SP FIRE 105: EXTERNAL WALL ASSEMBLIES AND FACADE CLADDINGS REACTION TO FIRE". Borås, Sweden.: Swedish National Testing and Research Institute Fire Technology; 1994.

106. Chinese Standards. (2012) "GB/T 29416-2012 Test method for fire-resistant performance of external wall insulation systems applied to building facades". China: Chinese Standards(GB); 2012.

107. Smolka, M., Messerschmidt, B., Scott, J., Madec, B.l., editors. (2013) "Semi-natural test methods to evaluate fire safety of wall claddings". 1st International Seminar for Fire Safety of Facades; 2013 14-15 November 2013; Paris, France: MATEC Web Conferences. Available from: http://www.matec-conferences.org/index.php?option=com_toc&url=/articles/matec-conf/abs/2013/07/contents/contents.html.

108. (1970) "LEPIR II. test method for fire spread".

109. Hungarian Standardisation Institute. (2009) "MSZ 14800-6 Fire resistance tests. Part 6: Fire propagation test for building facades". 2009.

110. Austrian Standards Institute. (2003) "Önorm B 3800-5 (draft) Fire behaviour of building materials and components—Part 5: Fire behaviour of facades—Requirements, tests and evaluations.". 2003.

111. (2008) "GOST 31251-2008. Facades of buildings. Fire hazard test method.". Interstate council for standardisation, metrology and certification; 2008.

112. ASTM. (1992) "Proposed Standard Test Method for Surface Flammability of Combustible Claddings and Exterior Wall Assemblies, ASTM Task Group E5.22.07 Vertical Channel Test". ASTM Draft, December 1992.

113. Whiting, P.N. (2005) "Development of the Vertical Channel Test Method for Regulatory Control Of Combustible Exterior Cladding Systems". New Zealand: Document No.: STUDY REPORT No. 137. Available from: http://www.branz.co.nz/cms_show_download.php?id=97d40985c67950e27bb801b212893cdec78fb228.

114. Nam, S., Bill, R.G. (2009) "A New Intermediate-scale Fire Test for Evaluating Building Material Flammability". Journal of Fire Protection Engineering. 2009 August 1, 2009;19(3):157-76.

115. Nam, S. (2007) "Intermediate-Scale Fire Test? Stepping Stone For Prediction Of Material Flamability In Real-Scale Fire Through Bench-Scale Fire Test Data". IAFSS AOFST7. 2007 (3).

116. Standards Australia. (2003) "Australian Standard ISO 9705—Fire Tests Full Scale Room Test for Surface Products". Sydney: Standards Association of Australia, 2003. Report No.: AS/ISO 9705.

117. NFPA. (2011) "NFPA 286: Standard Methods of Fire Tests for Evaluating Contribution of Wall and Ceiling Interior Finish to Room Fire Growth". Quincy, MA, USA: National Fire Protection Association.

118. International Conference of Building Officials. (1997) "UBC 26-3 Room Fire Test Standard for Interior of Foam Plastic Systems". International Conference of Building Officials, 1997.

119. ISO. (2002) "ISO 13784-1:2014: Reaction to fire test for sandwich panel building systems—Part 1: Small room test". Geneva, Switzerland: International Organization for Standardization; 2002.

120. ISO. (2002) "ISO 13784-2:2002:Reaction-to-fire tests for sandwich panel building systems—Part 2: Test method for large rooms". Geneva, Switzerland: International Organization for Standardization; 2002.

121. LPCB. (2005) "LPS 1181: PART 1: ISSUE 1.1, Series of Fire Growth Tests for LPCB Approval and Listing of Construction Product Systems Part One: Requirements and Tests for Built-up Cladding and Sandwich Panel Systems for Use as the External Envelope of Buildings". UK.: Loss Prevention Certification Board. Available from: http://www.redbook-live.com/pdf/LPS1181_part_one_1.1.pdf.

122. LPCB. (2005) "LPS 1181: PART 2: ISSUE 2.0, Series of Fire Growth Tests for LPCB Approval and Listing of Construction Product Systems Part Two: Requirements and tests for sandwich panels and built-up systems for use as internal constructions in buildings". UK.: Loss Prevention Cerification Board. Available from: http://www.redbooklive.com/pdf/LPS1181_part_one_1.1.pdf.

123. ISO. (2010) "ISO 1182:2010 Reaction to fire tests for products—Non-combustibility test". Geneva, Switzerland: International Organization for Standardization; 2010.

124. BSI. (1970) "BS 476-4:1970 Fire tests on building materials and structures Non-combustibility test for materials". UK.: British Standards Institute; 1970.

125. ASTM. (2012) "ASTM E136-12: Standard Test Method for Behavior of Materials in a Vertical Tube Furnace at 750°C". West Conshohocken, PA, United States: ASTM International; 2012.

126. Standards Australia. (1994) "AS 1530.1-1994: Methods for fire tests on building materials, components and structures—Combustibility test for materials". Sydney, Australia: SAI Global; 1994.

127. ASTM. (2012) "ASTM E2652-12 Standard Test Method for Behavior of Materials in a Tube Furnace with a Cone-shaped Airflow Stabilizer, at 750°C". West Conshohocken, PA, USA: ASTM International, 2012.

128. Babrauskas, V. (1992) "Chapter 4: The Cone Calorimeter". Heat Release in Fires. Essex, England: Elsevier Science Ltd.

129. ASTM. (2004) "ASTM E 1354-04a Standrad Test Method for Heat and Visible Smoke Release Rates for Materials and Products Using an Oxygen Consumption Calorimeter". West Conshohocken, PA, USA: ASTM International, 2004.

130. International Organisation for Standardisation. (1993) "Fire Tests—Reaction to fire—Rate of heat release from building products (Cone calorimeter method)". Geneva: International Organisation for Standardisation, 1993. Report No.: ISO 5660-1.

131. Standards Australia. (1998) "Australian Standard 3837—Method of test for heat and smoke release rates for materials and products using an oxygen consumption calorimeter". Standards Australia, 1998. Report No.: AS/NZS 3837:1998.

132. ISO. (2010) "ISO 1716:2010 Reaction to fire tests for products—Determination of the gross heat of combustion (calorific value)". Geneva, Switzerland: International Organization for Standardization; 2010.

133. CEN. (2010) "CSN EN 13823: Reaction to fire tests for building products—Building products excluding floorings exposed to the thermal attack by a single burning item". European Committee for Standardization; 2010.

134. (2014) "SBI Image from FireSERT facilities web page": FireSERT, University of Ulster; 2014 [cited 2014 18 March 2014]. Available from: http://www.firesert.ulster.ac.uk/facilities.php.

135. ISO. (2010) "ISO 11925-2:2010: Reaction to fire tests—Ignitability of products subjected to direct impingement of flame—Part 2: Single-flame source test". Geneva, Switzerland: International Organization for Standardization; 2010.

136. BSI. (1997) "BS 476-7:1997 Fire tests on building materials and structures Method of test to determine the classification of the surface spread of flame of products". UK.: British Standards Institute; 1997.

137. NFPA. (2012) "NFPA 268: Standard Test Method for Determining Ignitability of Exterior Wall Assemblies Using a Radiant Heat Energy Source". Quincy, MA, USA: National Fire Protection Association.

138. NFPA. (2006) "NFPA 255: Standard Method of Test of Surface Burning Characteristics of Building Materials". Quincy, MA, USA: National Fire Protection Association.

139. NFPA. (2013) "NFPA 259: Standard Test Method for Potential Heat of Building Materials". Quincy, MA, USA: National Fire Protection Association.

140. Babrauskas, V. (1992) "Chapter 2: From Bunsen Burner to Heat Release Rate CalorimeterHeat Release in Fires". Heat Release in Fires. Essex, England: Elsevier Science Publishers Ltd.

141. ASTM. (2010) "ASTM D635-10: Standard Test Method for Rate of Burning and/or Extent and Time of Burning of Plastics in a Horizontal Position". West Conshohocken, PA, United States: ASTM International; 2010.

142. ASTM. (2010) "ASTM E2307-10: Standard Test Method for Determining Fire Resistance of Perimeter Fire Barrier Systems Using Intermediate-Scale, Multi-story Test Apparatus". West Conshohocken, PA, United States: ASTM International; 2010.

143. CEN. (2014) "BS EN 1364-3:2014 Fire resistance tests for non-loadbearing elements. Curtain walling. Full configuration (complete assembly)". European Committee for Standardization; 2014.

144. CEN. (2014) "BS EN 1364-4:2014 Fire resistance tests for non-loadbearing elements. Curtain walling. Part configuration". European Committee for Standardization; 2014.

Printed in the United States
By Bookmasters